Automotive Technology

Automotive Technology

Todd Stephens

Larsen & Keller
www.larsen-keller.com

Automotive Technology
Todd Stephens
ISBN: 978-1-64172-663-4 (Hardback)

© 2022 Larsen & Keller

 Larsen & Keller

Published by Larsen and Keller Education,
5 Penn Plaza,
19th Floor,
New York, NY 10001, USA

Cataloging-in-Publication Data

Automotive technology / Todd Stephens.
    p. cm.
Includes bibliographical references and index.
ISBN 978-1-64172-663-4
1. Automobiles--Maintenance and repair. 2. Automobiles--Design and construction.
3. Motor vehicles--Design and construction. I. Stephens, Todd.
TL152 .A98 2022
629.2--dc23

For more information regarding Larsen and Keller Education and its products, please visit the publisher's website www.larsen-keller.com

TABLE OF CONTENTS

PREFACE

This book is a culmination of my many years of practice in this field. I attribute the success of this book to my support group. I would like to thank my parents who have showered me with unconditional love and support and my peers and professors for their constant guidance.

The practical application of the knowledge of self-propelled machines or vehicles is known as automotive technology. The study focuses on fuel and ignition systems, transmissions, engine construction, power trains, brakes, and electronic and diagnostic equipment. It also involves designing, construction, repairing and maintaining of all types of automobiles. There are several technologies, which are emerging at a rapid pace such as vehicle safety technology. It is developed to ensure security and safety of the passengers and automobiles. Significant examples include theft deterrence, remote speed sensing, geo-fencing capabilities, vehicle-to-vehicle communication, damage mitigation, etc. Sustainable automotive air conditioning is the next generation air conditioning of automobile. Other technologies include automotive steering and suspension technologies, automatic manual transmission, etc. This book is a compilation of chapters that discuss the most vital concepts in the field of automotive technology. The various sub-fields of automotive technology along with its technological progress that have future implications are glanced at in this book. It will provide comprehensive knowledge to the readers.

The details of chapters are provided below for a progressive learning:

Chapter – Introduction

Automobile is a four wheeled motorized vehicle that is powered by an internal engine used for transportation. It can be classified on the basis of purpose, capacity, fuel source, transmission, etc. The topics elaborated in this chapter will help in gaining a better perspective about automobiles.

Chapter – Automobile Engine and its Types

Automobile engine helps in the movement of an automobile. The types of automobile engine can be categorized into flathead engine, two-stroke engine, H engine, pushrod engine, single cylinder engine, V engine, diesel engine, etc. This chapter has been carefully written to provide an easy understanding of automobile engines and its types.

Chapter – Automotive Parts

Automotive parts include body, automotive valvetrain, suspension system, rolling chassis, backbone chassis, bumper and trunk in a car, automobile wheels, etc. This chapter closely examines these different components of automobiles to provide an extensive understanding of the subject.

Chapter – Automobile Systems

Automobile systems consist of independent parts that are capable of functioning by itself. Its major system includes engine lubrication system, cooling system, electrical system, transmission system, etc. This chapter delves into these major automobile systems to provide an in-depth understanding of the subject.

Chapter – Automobile Safety

Automobile safety deals with the study of design, construction and regulations for minimizing the occurrence and consequences of traffic accidents. Crashworthiness, global NCAP crash test, rollover, active safety, etc. are some concepts that fall within it. This chapter discusses these concepts of automobile safety in detail.

Todd Stephens

Introduction

Automobile is a four wheeled motorized vehicle that is powered by an internal engine used for transportation. It can be classified on the basis of purpose, capacity, fuel source, transmission, etc. The topics elaborated in this chapter will help in gaining a better perspective about automobiles.

AUTOMOBILES

An automobile can be defined as "A self-propelled passenger vehicle, used for land transport and generally has four to eight tires, and is powered by an internal combustion engine or an electric motor".

The branch of Engineering which deals with the manufacturing and technology of automotive vehicles is known as Automobile engineering.

Automobiles are used for both passenger and goods transport, hence performing as a lifeline to humans.

There are several bases of classification of automobiles, hence they can be categorized as follows:

Based on Purpose

- Passenger vehicle.
- Goods vehicle.

Based on Capacity

- Heavy motor vehicle (HMV).
- Medium motor vehicle (MMV).
- Light motor vehicle (LMV).

Based on Fuel Used

- Petrol engine vehicles.
- Diesel engine vehicle.
- Gas vehicle.
- Solar vehicle.

- Hydrogen engine vehicle.
- Electric powered vehicle.
- Steam engine vehicles.
- Hybrid vehicles.
- Hybrid electric vehicle (HEV).

Based on Transmission

- Automatic transmission vehicle.
- Manual (or conventional) transmission vehicle.
- Semi-automatic transmission vehicle.

Based on Number of Wheels

- Two-wheeler.
- Three-wheeler.
- Four-wheeler.
- Six-wheeler and more.

Based on Driving Side

- Left handed drive automobile.
- Right handed drive automobile.

Based on Drive Availability

- Four-wheel drive.
- All wheel drive.

Components of an Engine

Engine is one of the most important parts of an automobile. It is the source of power to the vehicle which is used to propel it. It is very important to understand different parts of an engine.

Parts on an engine can be classified into two parts.

- Stationary or structure forming components.
- Moving or mechanism forming components.

Mechanism Forming Components

- Piston.

- Piston rings.
- Gudgeon pin.
- Connecting rod.
- Crank.
- Crankshaft.
- Camshaft.
- Valves.
- Valve operating mechanism timing gears.
- Chain and sprocket.
- Belt and pulley.
- Flywheel.

Structural Components

- Cylinder block.
- Cylinder head.
- Gaskets.
- Cylinder liner.
- Crankcase.
- Inlet and outlet manifold.
- Oil pan.
- Resonator.
- Muffler (or silencer).
- Vibration damper.
- Bearing.
- Fasteners.
- Turbocharger.

Resistance to the Motion of an Automobile

While moving a vehicle has to overcome several types of resistances offered to it. Broadly the resistance can be classified into following categories:

- Aerodynamic drag.
- Gradient resistance.

- Rolling resistance.
- Inertia.

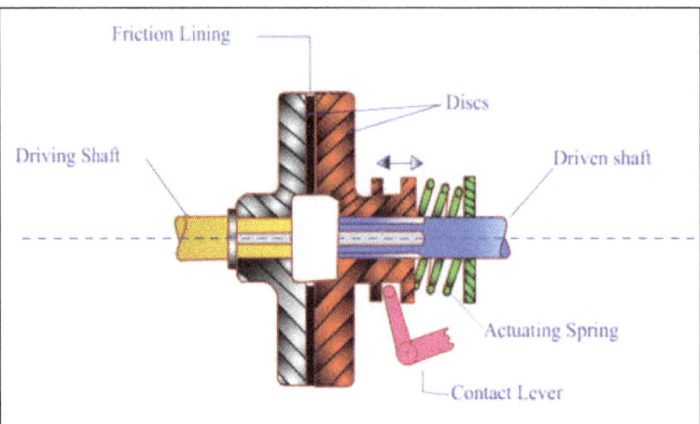

Clutch is a machine member which connects two rotating shafts, so that driven shaft can be stopped at the will of operator without stopping driving shaft.

A clutch has to provide interruptible connection between two shafts.

Clutches are used in automobiles and often used while changing gears or using brakes. In a modern automobile clutch is situated just between engine and gear box.

Types of Clutch

- Single plate clutch.
- Multi plate clutch.
- Centrifugal clutch.
- Semi centrifugal clutch.
- Cone clutch.

Clutch Operating Mechanisms

- Mechanically operated clutch.
- Hydraulically operated clutch.

Overdrive

Overdrive is a mechanical component which provides faster output speed for slower input speed.

Fluid Coupling

A fluid coupling is used to transmit motion between two shafts with the help of acceleration and deacceleration of working fluid. It works similar to the clutch.

Freewheel

It is a special type of clutch which disengages driven shaft from the driving shaft when the speed of driven shaft becomes more than driving shaft.

Gears

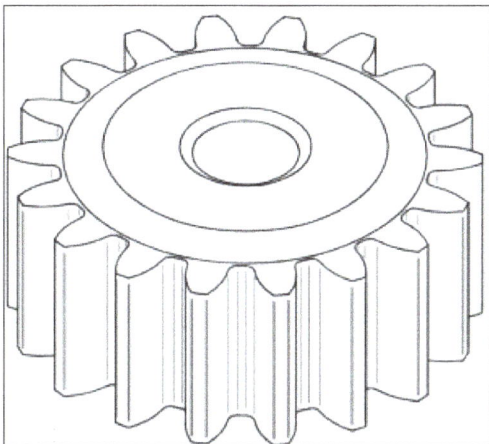

Gears are another important part of an automobile. These are used to adjust speed and torque of the wheels. Since high torque is required for starting a vehicle and high speed is required for running it.

For a given power engine speed and torque are always inversely proportional.

Types of Transmissions

Several kinds of transmissions can be employed in a vehicle. Transmissions can be classified as follows:

Manual Transmission

- Sliding mesh gearbox.
- Constant mesh gearbox.

- Synchronous gearbox.

- Synchronous gearbox with overdrive.

Semi-automatic Transmission

- Electric controlled with overdrive.

- Electric controlled without overdrive.

- Fluid torque drive.

Automatic Drive

- Hydraulic drive.

- Torque converter drive.

Braking System

Brakes are used to stop or slow down the vehicle. When brakes are applied it creates a force against the motion of tires which ultimately stops the vehicle.

Requirements of an Efficient Braking System

There are some requirements of a good braking system. It should ensure following things.

- It must operate on least effort.

- It produces less heat during operation.

- It should not slip.

- It must have strong braking force.

- It must act suddenly during emergency.

- The vehicle must stop at a smallest distance.

Types of Brakes

Following are the main types of the brakes used.

On the basis of Purpose Served

- Main brake.
- Parking brake.

On the basis of Location

- Wheel mounted.
- Transmission mounted.

On the basis of Driver's Ergonomics

- Hand brake.
- Foot brake.

On the basis of Actuating

- Mechanical brake.
- Hydraulic brake.
- Air brake.
- Electric brake.

On the basis of Construction

- Disc brake.
- Drum brake.

On the basis of Application of Brake Efforts

- Manual brake.
- Power brake.
- Power assisted.

On the basis of Action of Brake Shoes

- External expending brake.
- Internal expending brake.

Brake Lining

Drum shoes with linings.

Brake lining is a high friction material which is used to stop the moving vehicle by providing friction resistance to the motion of vehicle.

Properties of Brake Lining

For high performance brake lining should have following properties:

- Low heat swell.
- Low water swell.
- Low wear rate.
- High coefficient of hot and normal friction.
- High strength and physical properties.

Brake Lining Material

Brake linings are generally made from asbestos, rubber, metallic plates, resin, minerals and coefficient of friction modifiers.

Loads on the Frame

Frame of an automobile bears several types of loads. These loads are dependent of terrain and driving conditions.

1. Flexural (or bending) load.

Factors responsible for flexural load:

- Dead weight of vehicle.
- Weight of passengers.

- Engine torque.

- Braking torque.

- Road camber.

- Cornering force.

- Side wind.

2. Torsional load (or twisting moment).

Following things are employed to curb torsional load:

- Torque resisting members.

- Cross members.

- A radius rod.

- Benzo frame type torque members.

3. Impact load.

Steering System

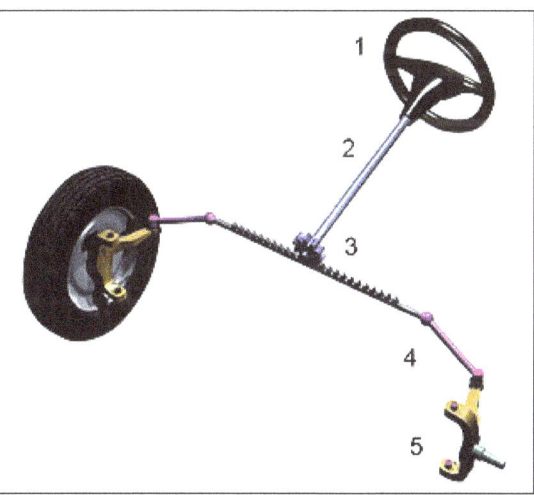

A steering system provides directional stability to a vehicle. It also helps in move the vehicle in a particular direction.

Types of Steering Systems

There are different types of steering systems. On the basis of gearing arrangement steering systems can be classified into following types:

- Worm and nut steering gear.

- Cam and lever steering gear.

- Worm and roller steering gear.

- Worm and sector steering gear.

- Rack and pinion steering gear

Differential Assembly

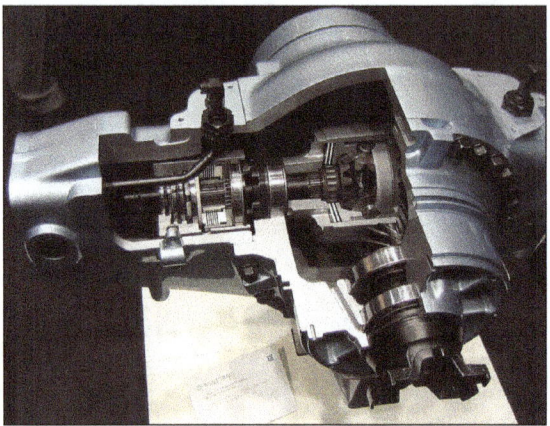

A differential assembly is a special kind of gear arrangement which allows outer tire to move with a greater speed than inner tires while turning.

Types of Differentials

- Double reduction type.

- Power lock or non-slip.

- Conventional.

Suspension System

A good suspension system is the luxury and necessity of a modern vehicle. It prevents and mitigates sudden jerks to the vehicle from the uneven road surface. A suspension system consists spring and damper elements for its working.

Types of Suspension Systems

- Leaf spring.
- Coil spring.
- Torsion bars.
- Air suspension.
- Automatic level control.

CLASSIFICATION OF AUTOMOBILES

Automobiles can be classified into several types based on several criteria. A brief classification of automobiles is listed below:

Based on purpose:

- Passenger vehicles – These automobiles carry passengers – e.g: Buses, Passenger trains, cars.
- Goods vehicles – These vehicles are used for transportation of goods from one place to another. e.g: Goods lorry, goods carrier.

Based on capacity:

- Heavy Motor Vehicle (HMV) – Large and bulky motor vehicles – e.g: Large trucks, buses.
- Light Motor Vehicle (LMV) – Small motor vehicles – e.g: Cars, Jeeps.
- Medium Vehicle – Relatively medium sized vehicles – e.g: Small trucks, mini buses.

Based on fuel source:

- Petrol engine vehicles – Automobiles powered by petrol engine – e.g: scooters, cars, mopeds, motorcycles.
- Diesel engine vehicles – Automotives powered by diesel engine – e.g: Trucks, Buses.
- Gas vehicles – Vehicles that use gas turbine as power source – e.g: Turbine powered cars.
- Solar vehicles – Vehicles significantly powered by solar power – e.g: Solar powered cars.
- Hydrogen vehicles – Vehicles that have hydrogen as a power source – e.g: Honda FCX Clarity.
- Electric vehicles – Automobiles that use electricity as a power source – e.g: Electric cars, electric buses.

- Steam Engine Vehicles – Automotives powered by steam engine – e.g: Steamboat, steam locomotive, steam wagon.

- Hybrid Vehicles – Vehicles that use two or more distinct power sources – e.g: Hybrid buses, hybrid cars like Toyota Prius, Honda Insight.

- Hybrid Electric Vehicle (HEV) – Automobile that uses both Internal Combustion Engine and Electric Power Source to propel itself – e.g: Jaguar C-X75.

Based on type of transmission:

- Automatic transmission vehicles – Automobiles that are capable of changing gear ratios automatically as they move – e.g: Automatic Transmission Cars.

- Conventional transmission vehicles – Automotives whose gear ratios have to be changed manually.

- Semi-automatic transmission vehicles – Vehicles that facilitate manual gear changing with clutch pedal.

Based on number of wheels:

- Two wheeler – Automobiles having two wheels – e.g: Scooters, motorcycles.

- Three wheeler – Automotive having three wheels – e.g: Tricycles, Auto rikshaws, Tempos.

- Four wheeler – Vehicle having four wheels – e.g: Car, Jeep.

- Six wheeler – Automobile having six wheels used for heavy transportation – e.g: Large trucks, large buses.

Based on the side of drive:

- Left hand drive automobile – Vehicle in which steering wheel is fitted on the left hand side – e.g: Automobiles found in USA, Russia.

- Right hand drive automobile - Vehicle in which steering wheel is fitted on the right hand side – e.g: Automobiles found in India, Australia.

PERFORMANCE OF AN AUTOMOBILE

To perform the car performance curve, we have to know the engine torque data at every operating speed. These data are informed by the engine performance curve. But if we do not have the engine curve, calculated simulation is needed. We can calculate the output torque and power from the engine, and then simulate the engine performance curve from the details of car specification; maximum power, maximum torque, and engine speed at these points as shown in equation:

$$T_i = 30 \frac{P_{max}}{\pi N_i} \left(A\left(\frac{N_i}{N_{Pmax}}\right) + B\left(\frac{N_i}{N_{Pmax}}\right)^2 - C\left(\frac{N_i}{N_{Pmax}}\right)^3 \right)$$

$$A = \frac{DE(2-E)-1}{E(2-E)-1}, B = \frac{1-A}{1-E/2},$$

$$C = \frac{BE}{2}, D = \frac{T_{max}}{T_{P\,max}}, E = \frac{N_{P\,max}}{N_{T\,max}}$$

Then, we use the engine torque and engine speed data to calculate with the transmission system and tire data to find the driving force and car velocity as shown in equation:

$$F_{ij} = \eta\frac{T_i i_j}{r}, V = 0.12\pi r\frac{N_i}{i_j}$$

Driving forces at each speed have to be reduced by resistances that is summarized from air resistance and rolling resistance. Air resistance is related to car square of velocity value, crosssection area, and drag coefficient of the car. Rolling resistance depends on the weight and rolling coefficient. The total resistance can be calculated as shown in equation:

$$R_{ij} = k_a A V_{ij}^2 + k_r W$$

After reducing the driving force by total resistance, we have the net force data. Car acceleration performance can be calculated from the net force and equivalence mass that is depended on gear position. The car acceleration can be calculated as shown in equation:

$$A_{ij} = \frac{F_{ij} - R_{ij}}{m_j}$$

$$m_j = \left(1.05 + 0.07 i_j^2\right) m$$

Finally, the overall car performance curve is represented by accelerate capability curve that presented correlation between accelerate performance related to the engine speed or car velocity.

Simulation Results and Discussions

For a better understanding about the concept to measure the overall car performance by the accelerate capability, we presented a case study simulated from specification data of Ford car; model Fiesta 5Dr 1.4L Style AT as shown in table.

Table: Car specification data.

Dimensions & Weight	
Overall Width (mm.)	1,722
Overall Height (mm.)	1,496
Weight (kg.)	1,127
Engine	
Maximum Power (kW/rpm)	70/5,750

Maximum Torque (Nm/rpm)	126/4,200
Transmission	
Gear Ratio 1st Gear	2.816
Gear Ratio 2nd Gear	1.498
Gear Ratio 3rd Gear	1.000
Gear Ratio 4th Gear	0.726
Final Gear Ratio	4.203
Tire Size	185/55 R15

Based on engine specification, we calculated output torque at engine speed from 600 to 7,200 rpm and set the speed range as 600 rpm. Transmission efficiency was assumed as 90% in calculating process. Simulated engine performance curve was shown in Figure.

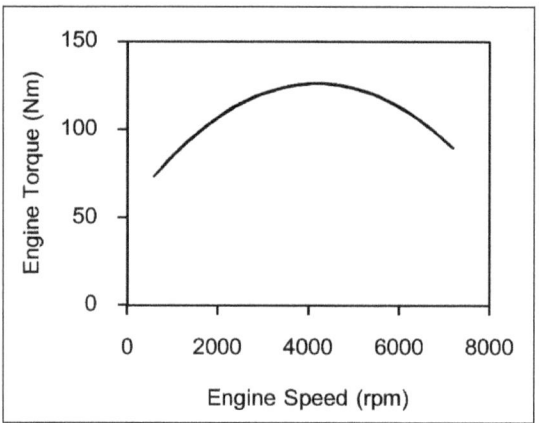

Simulated engine performance curve.

For the engine speed lower than 4,200 rpm, the engine output torque correlates with engine speed positively. The maximum torque is equal to 126 Nm at 4,200 rpm as shown in specification and decreases when the engine speed is over 4,200 rpm. However, this engine performance curve is not the exact data because it is calculated by mathematical simulation. It is always better if we have the information from the real performance curve.

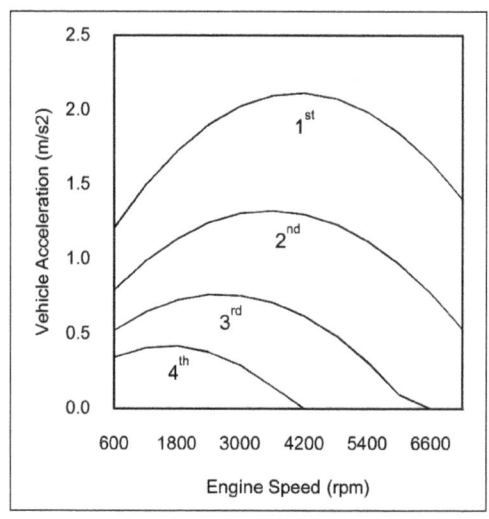

Simulated car performance curve.

Then, we used engine torque and operating speed results, with the tire radius of 292.25 mm. to calculate the driving force and car velocity. Total resistance was also analyzed by using assumption parameter by the following values (22); 0.80 for shape factor, 0.023 for air resistance coefficient, and 0.015 for rolling resistance coefficient. After simulating, we performed the car accelerate capability performance as contour plot between car acceleration (m/s2) and engine speed (rpm) at each gear position as shown in figure. From the figure, the areas under the curve line for each gear position were acceleration that the car can move at each gear position and not over the limit lines.

The overall car performance measured from acceleration capability is a fi nal result from overall parameters, such as engine output, transmission ratio, transmission efficiency, tire size, shape and car dimension, friction, electronic control unit, and driver skill. The concept to measure car performance from acceleration data is also feasible for an on-road experiment. Since currently, most cars have an electronic control unit (ECU), this concept is convenient to track digital input signals such as engine speed and car velocity to additional processor to analyze and display the result. Moreover, we can transfer raw data to process and display object on mobile equipment, such as a notebook PC, tablet PC, or smart phone.

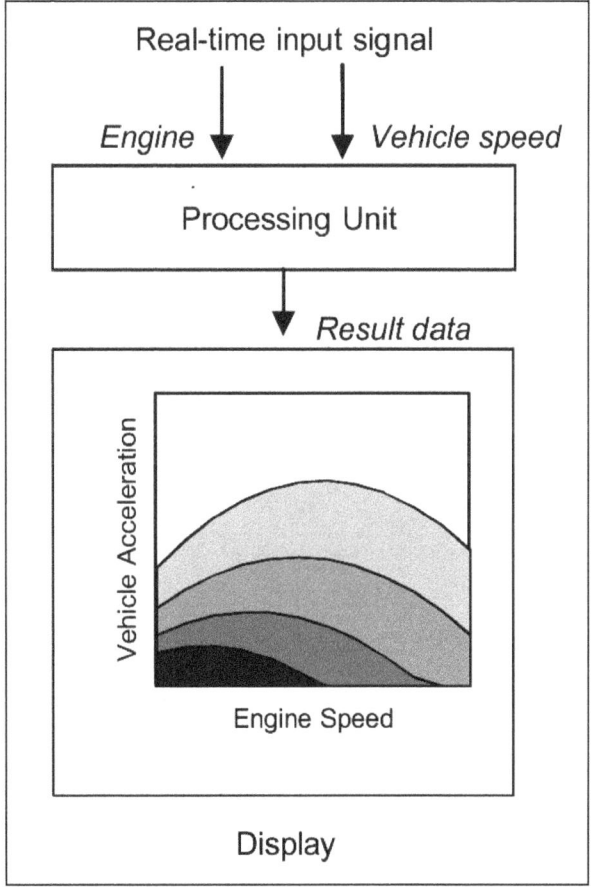

Conceptual implementation framework.

For future research, we will apply this concept to design and develop an equipment to collect digital input signals from existing ECU, to process the data, and to display the result following the conceptual implementation framework as shown in figure.

Automobile Engine and its Types

Automobile engine helps in the movement of an automobile. The types of automobile engine can be categorized into flathead engine, two-stroke engine, H engine, pushrod engine, single cylinder engine, V engine, diesel engine, etc. This chapter has been carefully written to provide an easy understanding of automobile engines and its types.

Here are the main parts of an automobile's engine:

Engine Block

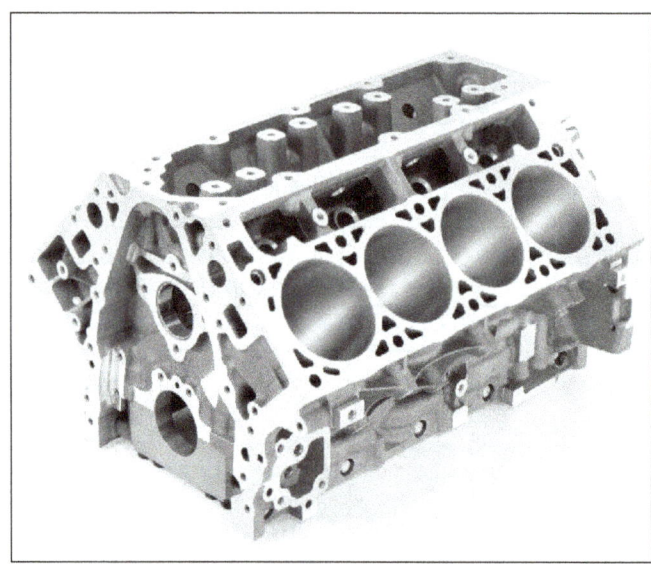

Engine block is an important parts of an engine. It is made by pouring the molten iron or aluminum alloy into a mold. The mold is made such that we should have required number of holes in the casted block, which are said to be the number of cylinders of an engine or engine cylinders. The diameter of these holes is called the bore of an engine.

We have some more holes along the length of engine cylinder, these are water and oil flow paths required for cooling and lubrication of an engine. Oil paths or vents are rather narrower than water flow vents.

What else we have in an engine block is semi-circular seats. On these seats half part of thrust bearings (thrust bearings come in two parts) are mounted, then we place crankshaft in these thrust bearings. But we still need to hold crankshaft to the engine block, to do that we have bearing-caps.

Bearing-caps have a semi-circular seat for other half part of thrust bearing. To mount the bearing-cap with engine block we use studs and nuts. One threaded end of stud goes into internal threaded hole in engine block, and other threaded end of stud goes into hole in bearing-cap and we fasten them together with nut. Two studs are used for holding one bearing-cap in place.

Now that we know that why we have semi-circular seats, in an engine block, let's take a look what we going to do with cylindrical holes in the block.

Piston

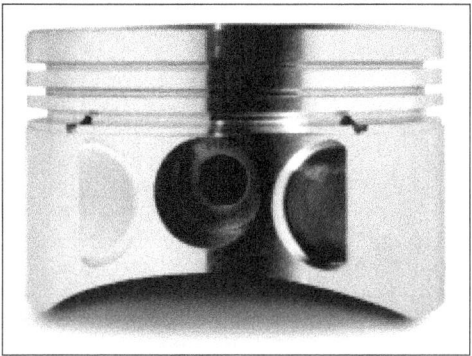

Piston is a cylindrical structure with a flat surface called crown at the top. Piston is the component that moves up and down in an engine cylinder. Wait what it would cause?, friction if one cylinder (piston) moving up and down in another cylinder? Yeah to overcome this problem groves are made on the circumference of this cylindrical structure (piston). And we place rings in these groves called piston rings. So now the whole cylindrical structure is not rubbing with engine cylinder and only piston rings are in contact with engine cylinder thus reducing friction to a great extent.

Now how we go to use this up and down motion of piston, for that we need to know about 2 more things that are connecting rod and wrist pin.

Connecting Rod

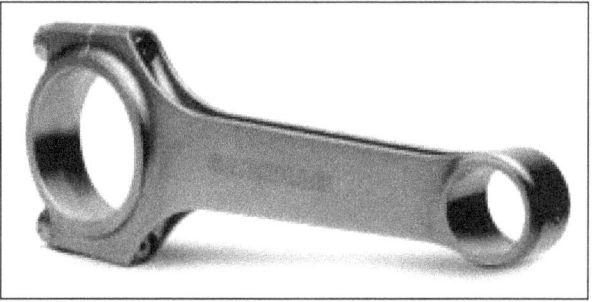

It is an 'I' shape structure whose one end is connected to piston and other one to crankshaft. The piston side end of connecting rod has hole in it. And we have also got a hole in piston's cylindrical structure just beneath the piston rings. So we align this hole with connecting rods hole and put a wrist pin through it. Wrist pin act as a bearing and connecting rod can move like pendulum beneath piston, though piston's cylindrical structure is gonna limit its motion. To make sure that wrist pin should not move from its position it is restricted by snap ring at both sides.

Other end of the connecting rod can be split into two parts. First is semi-circular journal bearing seat, which is placed over crankshaft after installing the half part of journal bearing, in the seat. Other half is journal bearing cap. These two parts are bolted together, holding crankshaft in between. Thus piston is now connected to crankshaft through connecting rod.

Crankshaft

As name suggests it is designed in such a way to convert linear (up and down) motion of piston into rotational motion. It works same as slider- crank mechanism. Material used for making crankshaft is cast iron generally but we also use forged steel in high power engines where load on crankshaft is too high.

Casting a crank-shaft seems to be an easy task, but it's not. Once crankshaft is casted it is then machined, which is not that easy considering its shape. Then after machining it requires proper balancing to work properly.

You will generally find some random holes in crankshaft; these holes are for balancing the crankshaft while rotating at high speed.

Crankshaft Casing or Oil Sump

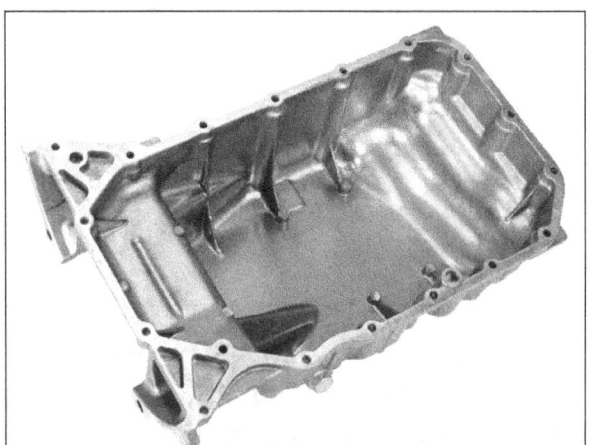

It is also called as oil sump. It is a casing which is bolted to engine block, which covers engine from bottom thus called crankshaft casing. It retains lubricating oil in it which is pumped to different engine parts. Crankshaft has got small holes which spills oil towards piston, to remove piston heat and lubricate the piston rings, so it also prevents oil from splashing. We have got a bolt at the bottom of this casing from where we remove used lubricating oil during maintenance.

Engine Head

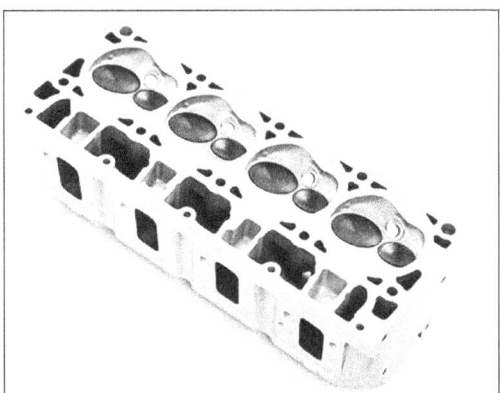

Engine head is casted in the same way as engine block. Its mold is made such that the casted piece must have an opening for air to flow into engine cylinder and an exhaust opening from where the burnt gases will go out. This passage of air flowing in and going out of engine cylinder is controlled by inlet and outlet valves. So engine head also has cylindrical holes to insert valve stem. Furthermore to burn the air-fuel mixture we have to fire it up and how we going to do that? We need a spark plug which must produce a spark inside engine cylinder, for this we need cylindrical hole in engine block to put spark plug into engine cylinder. We also have semi-circular seats casted in engine head for camshaft bearings.

We have 4 internally threaded holes at the top of engine block. Threaded end of the stud gets fastened up in engine block and we have 4 studs fasten up in the same way in the engine block, then we put a gasket whose holes align with the engine block studs. We have 4 holes through the engine head and they align with 4 fastened studs of engine block. So this way we fasten the engine block and engine head with gasket in between together with help of stud and nut assembly.

Valves

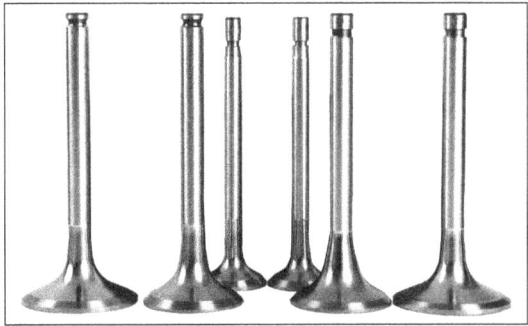

As we have already know that they control the inlet and exhaust air to go into and out of engine cylinder. Material used to make valves is nickel-chromium iron alloy. It can resist high temperature and have great strength. Valve could be described in two parts- valve stem and valve head. As we already know that we have got cylindrical hole in engine head for valve steam and we also have a valve seat where valve head will rest in engine head. Valve is mounted upside down means valve head is facing engine cylinder. It is so because when there would be high pressure in engine cylinder it would press the valve head against its seat in engine head and thus pressure will be maintained at best.

Camshaft

It is a shaft with a number of cam profiles along its length. So it regulates the valves opening and closing time. It does so by pressing the end of valve stem by its cam profile. But we still need a mechanism which would return the valve back to its position once pressed by the cam profile of camshaft. We have valve spring and bucket head tappet assembly for solving this problem.

Now as we have semicircular seats for bearings of camshaft. What retains it to be fixed in engine head while rotating is cam-caps? They hold the other half of journal bearing and have two holes in their casing through which we insert bolts and fasten them in internally threaded holes of engine head thus we hold our camshaft between the casted journal seat in engine block and cam-caps and fasten them up with long bolt.

Valve Spring and Tappet

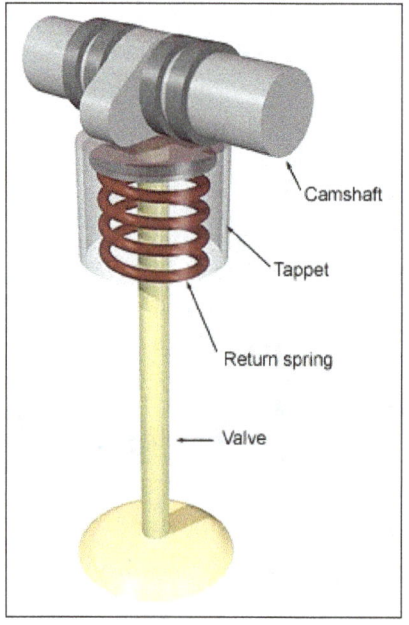

Valve spring provides a self-returning mechanism when valve is not being pressed by camshaft. We further have bucket type tappet covering the valve spring. The purpose of tappet is to provide smooth surface for cam to press the valve spring or inlet and outlet valve. The arrangement is like we have valve spring around valve stem and tappet mounted over that spring for smooth surface and camshaft mounted just over it making the valve move up and down using its cam-profile.

Timing Belt

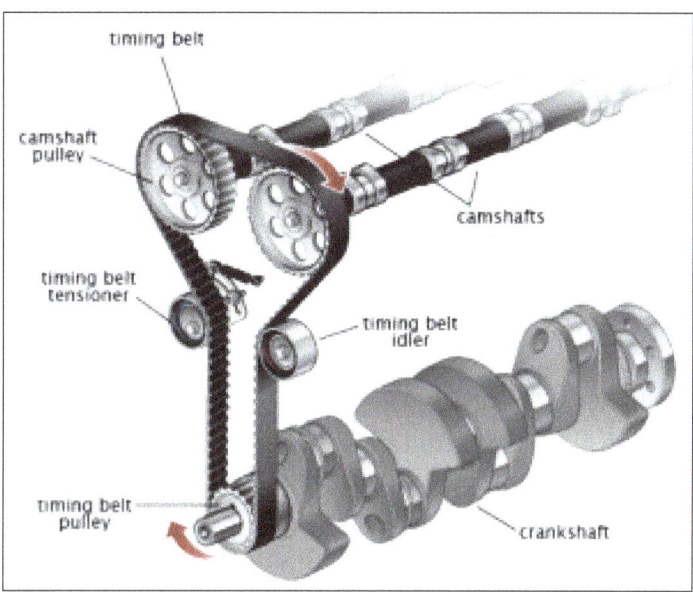

Wonder how camshaft gets its rotational motion to regulate valves. Yeah it's through timing belt which transmits the motion of gear mounted at crankshaft called crank-gear, to the gear mounted at camshaft. The ration of cam-gear to that of crank-gear is 2:1. So that camshaft would rotate only once in two rotations of crankshaft. Timing belt is made up either from glass-fiber or Kevlar so it does not worn-out easily.

Spark Plug

sparkplug

It is the parts of an engine that ignites the air-fuel mixture in the engine cylinder. It produces the spark at right time by using electrical energy of battery. The basic working principle is that when we have high electrical potential at one end and zero or negative potential at other end. And two ends being real very close to each other, it produces an electric field so strong between them that it ionizes the air molecules thus producing spark. And that is in the combustion chamber, It is made up of titanium so that it can withstand a real high temperature generated by high electric potential difference while producing spark.

Gasket

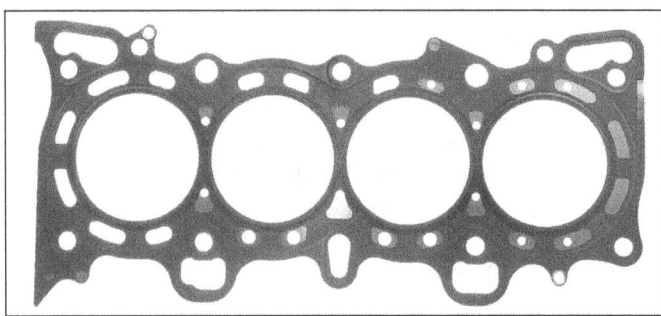

A wide variety of materials are used in making gaskets like Teflon, glass-fiber, silicon etc. It is generally a paper like sheet which is placed between engine block and engine head. we have both water and oil vents in engine block, so gasket gives insulation from water or oil leaking into engine cylinder or air-fuel mixture from engine cylinder leaking out from joint of engine block and engine head. Aluminum engine blocks are preferred over cast iron because it expends more on heating thus compressing the gasket more, increases the workability of gasket, thus reducing the chances of leakage.

Piston Rings

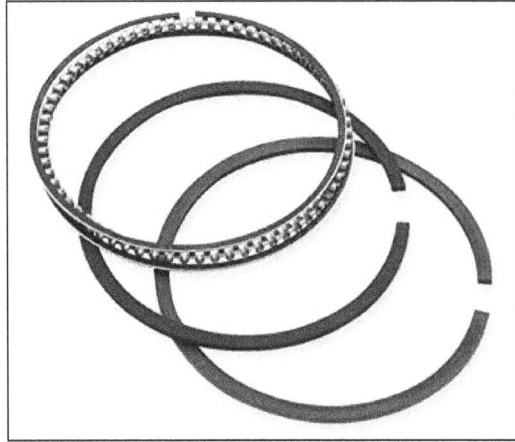

We have talked about them; they reduce friction between piston and cylinder walls. What else do redy they do?

Piston rings prevent the pressure created by burning of air-fuel mixture from leaking into crank-case. Not only that piston rings scrap down the oil from cylinder walls which is spilled by crank-shaft to remove heat from piston. They also transfer heat of the piston to the cylinder walls which are being cooled by water circulation through water vents.

FLAT ENGINE

The 'Boxer' engine is so called because the movement of the engine's pistons resemble the movement of a boxer's fists in the horizontal plane. However, unlike the boxer's fists that both move in the same direction, half of the 'Boxer' engine's pistons move in the opposite direction.

This very important difference provides the key benefits of the horizontally-opposed (Boxer) engine layout where half of the total number of cylinders and therefore pistons lay on their sides in an east-west configuration moving in opposite directions.

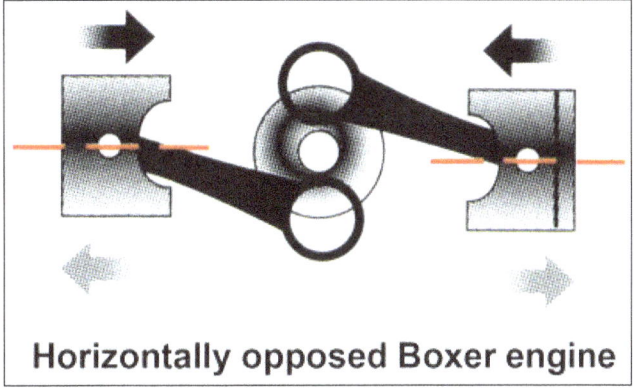

Horizontally opposed Boxer engine

Contribution to Vehicle Handling and Stability

The key benefit of the Subaru 'Boxer' engine's horizontally-opposed layout is not about its power output or environmental performance but its contribution to the vehicle's handling, stability and as a result, safety.

Firstly, because the cylinders in the Boxer engine lay on their sides the overall height of the engine is significantly lower especially when compared to the more normal 'inline' engine where all the pistons move in the same direction in a north-south configuration. This results in a low centre of gravity that has very significant advantages in terms of the vehicle's handling and stability, by helping to keep the tyres more firmly planted on the road surface as a result of a lower level of weight transfer from the inside wheel to the outside wheel when cornering.

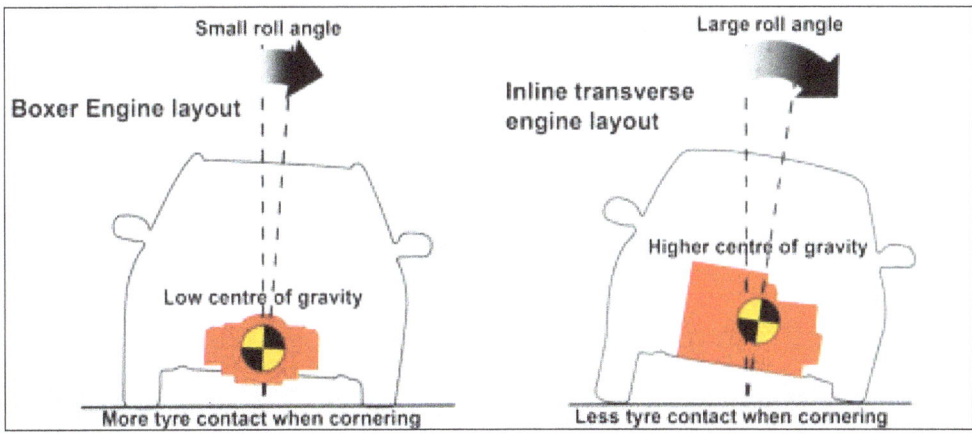

Furthermore, because half of the total number of cylinders is on one side and the other half on the opposite side with a common crankshaft sandwiched in the middle the left-right weight distribution with the engine positioned on the centre line of the vehicle is equal. This also makes a very significant contribution to vehicle balance, stability and handling particularly when cornering or making rapid changes of direction.

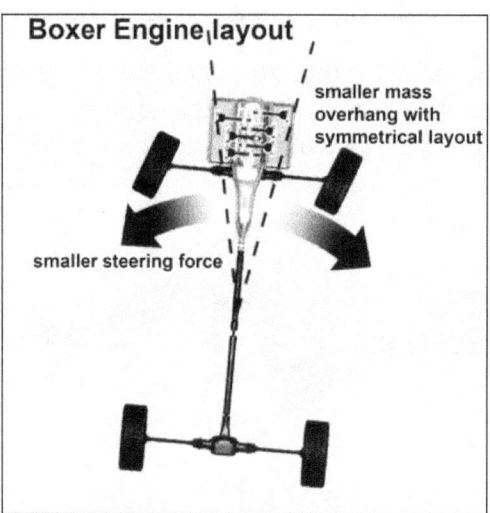

The overall length of the Boxer engine is comparatively short compared to a conventional 'in line' or 'V' engine layout as a result of the degree of cylinder overlap that is possible due to the opposed piston configuration. This contributes to an improvement in vehicle front-rear weight distribution that again is a significant factor in vehicle handling, stability and also steering response. If a vehicle's weight is more centrally positioned the forces required to change direction are less due to a lower level of weight positioned in front of the steering axle. The result is more responsive and precise steering, again a key safety attribute.

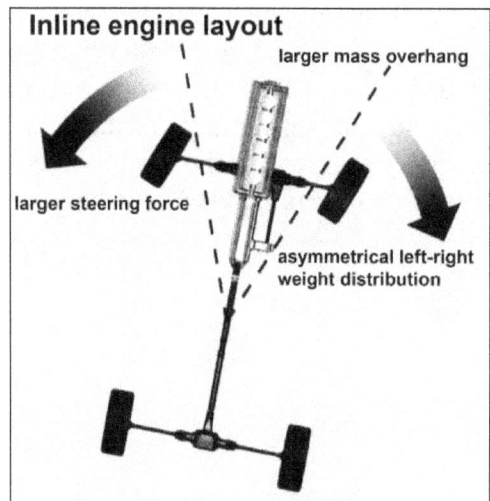

Durability, Reliability and Lightweight

Because the cylinders lay on their sides positioned in opposite directions with a common crankshaft, the size and therefore the weight of the Boxer engine is relatively lightweight. Subaru 'Boxer' engines also feature all aluminium construction of the crankcase and cylinder heads that also significantly contributes to weight reduction. This not only improves the power to weight ratio and therefore driving and environmental performance but also again helps handling and stability.

Because the pistons move in opposite directions the natural rotational balance of the engine is very good due to the cancelation of the piston inertia forces that move in opposite directions. This means two things:

- Good rotational balance means lower levels of vibration and harshness, which translates into less wear and tear and therefore higher levels of durability meaning a more reliable and lower cost of ownership.

- Because the crankshaft is sandwiched between two very rigid crankcase halves it is very well supported and therefore its durability is significantly enhanced and because it absorbs less vibration its design can be of a much lighter construction. This not only contributes to the over lightness of the engine but also means a freer revving more responsive engine, both attributes bringing improvements in vehicle active safety through improved driveability.

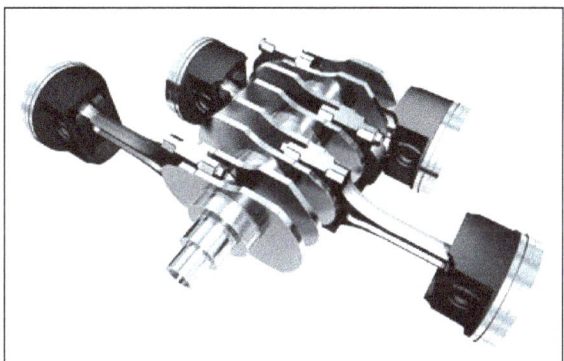

Crash Safety Contribution

Another key benefit of the Subaru 'Boxer' engine design is that it could save your life! This is the result of its low height when positioned in the front engine bay. In a full frontal or even an offset accident a severe impact will cause the front of the vehicle to crumple and subsequently cause the engine to be driven backwards towards the passenger compartment with potential penetration and severe trauma to the front occupants. Due to its low height, the 'Boxer' engine can be redirected under the floor of the passenger compartment avoiding any penetration of the passenger compartment and therefore reducing the chance of front occupant injury.

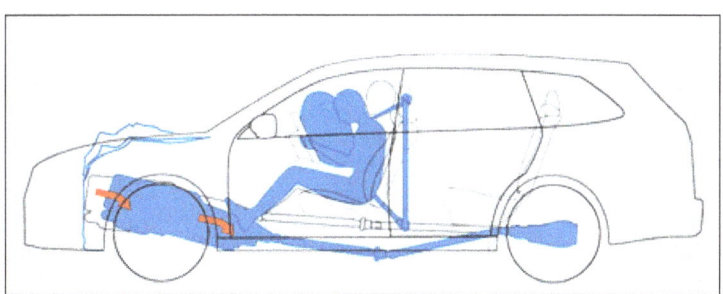

Disadvanatges of a Boxer Engine

What are the negative aspects of the 'Boxer' horizontally opposed engine design? Why do all manufacturers not adopt this format if it has so many advantages?

Traditionally one of the negative aspects of 'Boxer' engine design was that as the demand for bigger capacity engines increased the only way to achieve a bigger engine was through an increase in the cylinder bore (diameter) size. This was because any increase in the length of the cylinder (stroke) would make the overall engine width greater. This meant difficulty in accommodating the wider engine within the chassis without making the overall width of the car greater, which is not desirable from a packaging point of view and given our traffic and road conditions.

When an engine becomes 'over square' meaning the bore is greater than the stroke, the efficiency of combustion becomes increasingly difficult. Given the minimal time that is available in a relatively high revving engine on each cycle to completely burn all of the fuel, if the bore size becomes too large the flame travel time from the spark plug to the outside of the cylinder wall becomes problematic. The result is that the fuel is incompletely burnt and fuel consumption deteriorates. Improvements in combustion chamber design and computer control of the air-fuel ratio, ignition and valve timing has been able to successfully manage this slightly negative attribute of the 'Boxer' engine design.

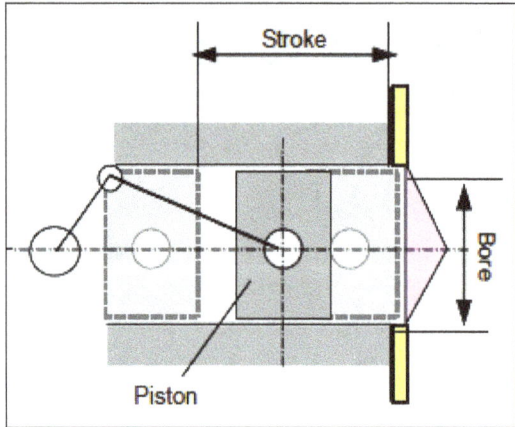

However, the launch of the third generation Subaru 'Boxer' engine has seen the introduction of a new innovative design that has allowed an increase in the engine stroke and reduction of the bore size, without increase in the overall engine width, has overcome this previously negative design feature of the 'Boxer' engine.

The other negative aspect of the 'Boxer' engine design is its relatively complex shape that causes an increase in manufacturing costs.

FLATHEAD ENGINE

A flathead engine, otherwise **sidevalve engine,** is an internal combustion engine with its poppet valves contained within the engine block, instead of in the cylinder head, as in an overhead valve engine.

Flatheads are an early design concept that has mostly fallen into disuse, but they are currently experiencing a revival in low-revving aero-engines such as the D-Motor.

The Side-valve Design

The valve gear comprises a camshaft sited low in the cylinder block which operates the poppet valves via tappets and short pushrods (or sometimes with no pushrods at all). The flathead system obviates the need for further valvetrain components such as lengthy pushrods, rocker arms, overhead valves or overhead camshafts. The sidevalves are typically adjacent, sited on one side of the cylinder(s), though some flatheads employ the less common "crossflow" "T-head" variant. In a T-head engine, the exhaust gases leave on the opposite side of the cylinder from the intake valve.

The sidevalve engine's combustion chamber is not above the piston (as in an OHV (overhead valve) engine) but to the side, above the valves. The spark plug may be sited over the piston (as in an OHV engine) or above the valves; but aircraft designs with two plugs per cylinder may use either or both positions.

"Pop-up pistons" may be used with compatible heads to increase compression ratio and improve the combustion chamber's shape to prevent knocking. "Pop-up" pistons are so called because, at tdc, they protrude above the top of the cylinder block.

Advantages

The advantages of a sidevalve engine include: simplicity, reliability, low part count, low cost, low weight, compactness, responsive low-speed power, low mechanical engine noise, and insensitivity to low-octane fuel. The absence of a complicated valvetrain allows a compact engine that is cheap to manufacture, since the cylinder head may be little more than a simple metal casting. These advantages explain why side valve engines were used for economy cars, trucks, and agricultural engines for many years, while OHV designs came to be specified only for high-performance applications such as aircraft, luxury cars, sports cars, and some motorcycles.

At top dead centre, the piston gets very close to the flat portion of the cylinder head above, and the resultant squish turbulence produces excellent fuel/air mixing. A feature of the sidevalve design (particularly beneficial for an aero-engine) is that if a valve should seize in its guide and remain partially open, the piston would not be damaged, and the engine would continue operating safely on its other cylinders.

Disadvantages

The main disadvantages of a sidevalve engine are poor gas flow, poor combustion chamber shape, and low compression ratio, all of which result in a low-revving engine with low power output and low efficiency. Because sidevalve engines do not burn the fuel correctly, they suffer from high hydrocarbon emissions.

Sidevalve engines can only be used for engines operating on the Otto principle. The combustion chamber shape is unsuitable for Diesel engines.

In a sidevalve engine, intake and exhaust gases follow a circuitous route, with low volumetric efficiency, or "poor breathing", not least because the exhaust gases interfere with the incoming charge. Because the exhaust follows a lengthy path to leave the engine, there is a tendency for the engine to overheat. Although a sidevalve engine can safely operate at high speed, its volumetric efficiency swiftly deteriorates, so that high power outputs are not feasible at speed. High volumetric efficiency was less important for early cars because their engines rarely sustained extended high speeds, but designers seeking higher power outputs had to abandon the sidevalve. A compromise used by the Willys Jeep, Rover, Landrover, and Rolls-Royce in the 1950s was the "F-head" (or "intake-over-exhaust" valving), which has one sidevalve and one overhead valve per cylinder.

The flathead's elongated combustion chamber is prone to preignition (or "knocking") if compression ratio is increased, but improvements such as laser ignition or microwave enhanced ignition might help prevent knocking. Turbulence grooves may increase swirl inside the combustion chamber, thus increasing torque, especially at low rpm. Better mixing of the fuel/air charge improves combustion and helps to prevent knocking.

An advance in flathead technology resulted from experimentation in the 1920s by Sir Harry Ricardo, who improved their efficiency after studying the gas-flow characteristics of sidevalve engines.

The difficulty in designing a high-compression-ratio flathead means that most tend to be spark-ignition designs, and flathead diesels are virtually unknown.

TWO-STROKE ENGINE

A two-stroke (or two-cycle) engine is a type of internal combustion engine which completes a power cycle with two strokes (up and down movements) of the piston during only one crankshaft revolution. This is in contrast to a "four-stroke engine", which requires four strokes of the piston to complete a power cycle during two crankshaft revolutions. In a two-stroke engine, the end of the combustion stroke and the beginning of the compression stroke happen simultaneously, with the intake and exhaust (or scavenging) functions occurring at the same time.

Two-stroke engines often have a high power-to-weight ratio, power being available in a narrow range of rotational speeds called the "power band". Compared to four-stroke engines, two-stroke engines have a greatly reduced number of moving parts, and so can be more compact and significantly lighter.

Different Two-stroke Design Types

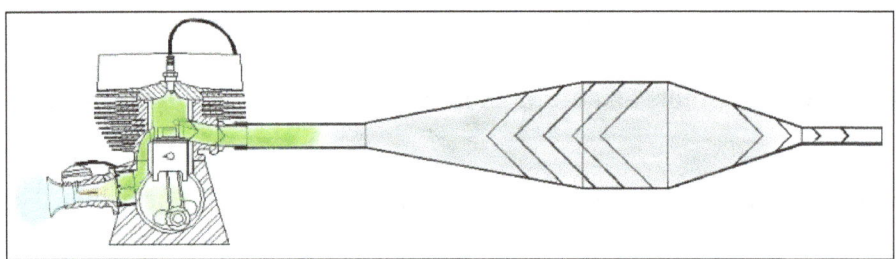

A two-stroke engine, in this case with an expansion chamber, illustrates the effect of a reflected pressure wave on the fuel charge. This is important for maximum charge pressure (volumetric efficiency) and fuel economy. It is used on most high-performance engine designs.

Although the principles remain the same, the mechanical details of various two-stroke engines differ depending on the type. The design types vary according to the method of introducing the charge to the cylinder, the method of scavenging the cylinder (exchanging burnt exhaust for fresh mixture) and the method of exhausting the cylinder.

Piston-controlled Inlet Port

Piston port is the simplest of the designs and the most common in small two-stroke engines. All functions are controlled solely by the piston covering and uncovering the ports as it moves up and down in the cylinder. In the 1970s, Yamaha worked out some basic principles for this system. They found that, in general, widening an exhaust port increases the power by the same amount as raising the port, but the power band does not narrow as it does when the port is raised. However, there is a mechanical limit to the width of a single exhaust port, at about 62% of the bore diameter for reasonable ring life. Beyond this, the rings will bulge into the exhaust port and wear quickly. A maximum 70% of bore width is possible in racing engines, where rings are changed every few races. Intake duration is between 120 and 160 degrees. Transfer port time is set at a minimum of 26 degrees. The strong low pressure pulse of a racing two-stroke expansion chamber can drop the pressure to -7 PSI when the piston is at bottom dead center, and the transfer ports nearly wide open. One of the reasons for high fuel consumption in two-strokes is that some of the incoming pressurized fuel-air mixture is forced across the top of the piston, where it has a cooling action, and straight out the exhaust pipe. An expansion chamber with a strong reverse pulse will stop this out-going flow. A fundamental difference from typical four-stroke engines is that the two-stroke's crankcase is sealed and forms part of the induction process in gasoline and hot bulb engines. Diesel two-strokes often add a Roots blower or piston pump for scavenging.

Reed Inlet Valve

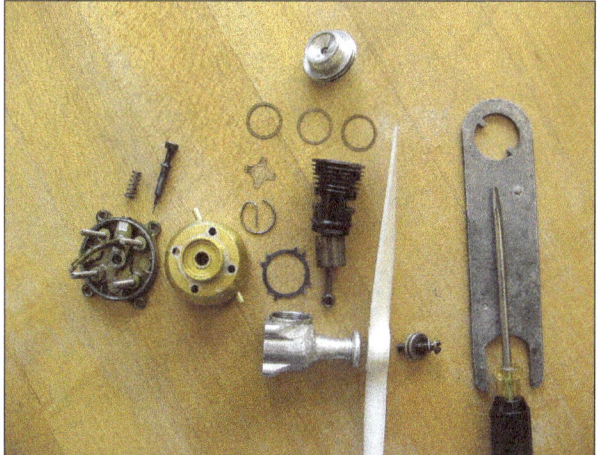

A Cox Babe Bee 0.049 cubic inch (0.8 cubic cm) reed valve engine, disassembled, uses glow plug ignition. The mass is 64 grams.

The reed valve is a simple but highly effective form of check valve commonly fitted in the intake tract of the piston-controlled port. They allow asymmetric intake of the fuel charge, improving power and economy, while widening the power band. They are widely used in motorcycle, ATV and marine outboard engines.

Rotary Inlet Valve

The intake pathway is opened and closed by a rotating member. A familiar type sometimes seen on small motorcycles is a slotted disk attached to the crankshaft which covers and uncovers an opening in the end of the crankcase, allowing charge to enter during one portion of the cycle (aka disc valve).

Another form of rotary inlet valve used on two-stroke engines employs two cylindrical members with suitable cutouts arranged to rotate one within the other - the inlet pipe having passage to the crankcase only when the two cutouts coincide. The crankshaft itself may form one of the members, as in most glow plug model engines. In another embodiment, the crank disc is arranged to be a close-clearance fit in the crankcase, and is provided with a cutout which lines up with an inlet passage in the crankcase wall at the appropriate time, as in Vespa motor scooters.

The advantage of a rotary valve is that it enables the two-stroke engine's intake timing to be asymmetrical, which is not possible with piston-port type engines. The piston-port type engine's intake timing opens and closes before and after top dead center at the same crank angle, making it symmetrical, whereas the rotary valve allows the opening to begin and close earlier.

Rotary valve engines can be tailored to deliver power over a wider speed range or higher power over a narrower speed range than either piston port or reed valve engine. Where a portion of the rotary valve is a portion of the crankcase itself, it is particularly important that no wear is allowed to take place.

Cross-flow-scavenging

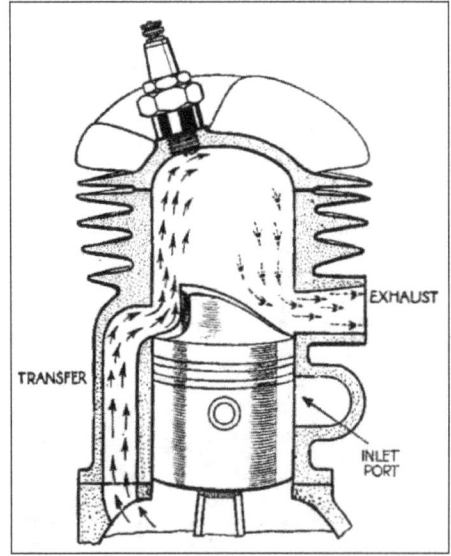

Deflector piston with cross-flow scavenging.

In a cross-flow engine, the transfer and exhaust ports are on opposite sides of the cylinder, and a deflector on the top of the piston directs the fresh intake charge into the upper part of the cylinder, pushing the residual exhaust gas down the other side of the deflector and out the exhaust port. The deflector increases the piston's weight and exposed surface area, affecting piston cooling and also making it difficult to achieve an efficient combustion chamber shape. This design has been superseded since the 1960s by the loop scavenging method (below), especially for motorbikes, although for smaller or slower engines, such as lawn mowers, the cross-flow-scavenged design can be an acceptable approach.

Loop-scavenging

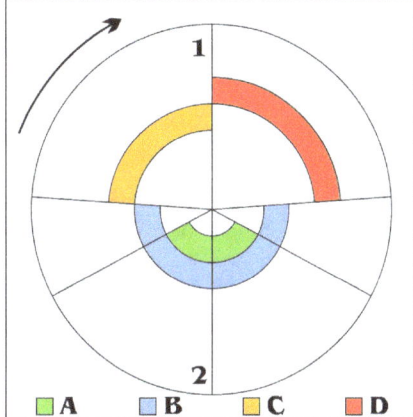

The two-stroke cycle: 1. Top dead center (TDC) and 2.Bottom dead center (BDC)
A: Intake/scavenging, B: Exhaust, C: Compression, D: Expansion (power).

This method of scavenging uses carefully shaped and positioned transfer ports to direct the flow of fresh mixture toward the combustion chamber as it enters the cylinder. The fuel/air mixture strikes the cylinder head, then follows the curvature of the combustion chamber, and then is deflected downward.

This not only prevents the fuel/air mixture from traveling directly out the exhaust port, but also creates a swirling turbulence which improves combustion efficiency, power and economy. Usually, a piston deflector is not required, so this approach has a distinct advantage over the cross-flow.

Often referred to as "Schnuerle" (or "Schnürle") loop scavenging after Adolf Schnürle, the German inventor of an early form in the mid-1920s, it became widely adopted in that country during the 1930s and spread further afield after World War II.

Loop scavenging is the most common type of fuel/air mixture transfer used on modern two-stroke engines. Suzuki was one of the first manufacturers outside of Europe to adopt loop-scavenged two-stroke engines. This operational feature was used in conjunction with the expansion chamber exhaust developed by German motorcycle manufacturer, MZ and Walter Kaaden.

Loop scavenging, disc valves and expansion chambers worked in a highly coordinated way to significantly increase the power output of two-stroke engines, particularly from the Japanese manufacturers Suzuki, Yamaha and Kawasaki. Suzuki and Yamaha enjoyed success in grand Prix motorcycle racing in the 1960s due in no small way to the increased power afforded by loop scavenging.

An additional benefit of loop scavenging was the piston could be made nearly flat or slightly dome shaped, which allowed the piston to be appreciably lighter and stronger, and consequently to tolerate higher engine speeds. The "flat top" piston also has better thermal properties and is less prone to uneven heating, expansion, piston seizures, dimensional changes and compression losses.

SAAB built 750 and 850 cc 3-cylinder engines based on a DKW design that proved reasonably successful employing loop charging. The original SAAB 92 had a two-cylinder engine of comparatively low efficiency. At cruising speed, reflected wave exhaust port blocking occurred at too low a frequency. Using the asymmetric three-port exhaust manifold employed in the identical DKW engine improved fuel economy.

The 750 cc standard engine produced 36 to 42 hp, depending on the model year. The Monte Carlo Rally variant, 750 cc (with a filled crankshaft for higher base compression), generated 65 hp. An 850 cc version was available in the 1966 SAAB Sport (a standard trim model in comparison to the deluxe trim of the Monte Carlo). Base compression comprises a portion of the overall compression ratio of a two-stroke engine. Work published at SAE in 2012 points that loop scavenging is under every circumstance more efficient than cross-flow scavenging.

Uniflow-scavenging

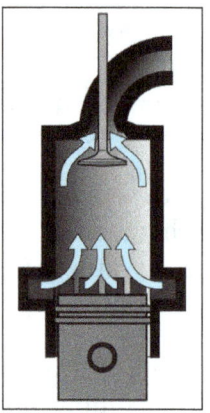

Uniflow scavenging.

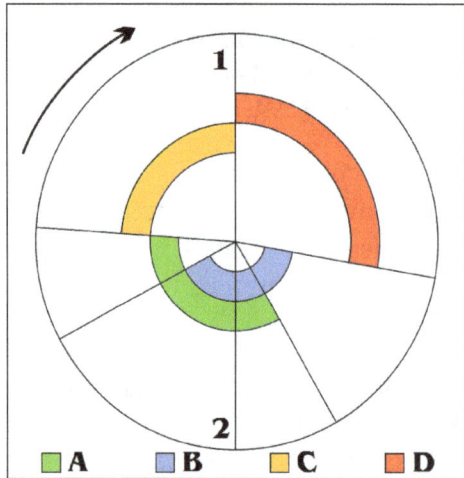

The uniflow two-stroke cycle: 1.Top dead center (TDC) and 2. Bottom dead center (BDC) A: Intake (effective scavenging, 135°–225°; necessarily symmetric about BDC; Diesel injection is usually initiated at 4° before TDC),B: Exhaust, C: Compression,D: Expansion (power).

In a uniflow engine, the mixture, or "charge air" in the case of a diesel, enters at one end of the cylinder controlled by the piston and the exhaust exits at the other end controlled by an exhaust valve or piston. The scavenging gas-flow is therefore in one direction only, hence the name uniflow. The valved arrangement is common in on-road, off-road and stationary two-stroke engines (Detroit Diesel), certain small marine two-stroke engines (Gray Marine), certain railroad two-stroke diesel locomotives (Electro-Motive Diesel) and large marine two-stroke main propulsion engines (Wärtsilä). Ported types are represented by the opposed piston design in which there are two pistons in each cylinder, working in opposite directions such as the Junkers Jumo 205 and Napier Deltic. The once-popular split-single design falls into this class, being effectively a folded uniflow. With advanced angle exhaust timing, uniflow engines can be supercharged with a crankshaft-driven (piston or Roots) blower.

Stepped Piston Engine

The piston of this engine is "top-hat" shaped; the upper section forms the regular cylinder, and the lower section performs a scavenging function. The units run in pairs, with the lower half of one piston charging an adjacent combustion chamber.

This system is still partially dependent on total loss lubrication (for the upper part of the piston), the other parts being sump lubricated with cleanliness and reliability benefits. The piston weight is only about 20% heavier than a loop-scavenged piston because skirt thicknesses can be less. Bernard Hooper Engineering Ltd. (BHE) is one of the more recent engine developers using this approach.

FOUR-STROKE ENGINE

A four-stroke (also four-cycle) engine is an internal combustion (IC) engine in which the piston completes four separate strokes while turning the crankshaft. A stroke refers to the full travel of the piston along the cylinder, in either direction. The four separate strokes are termed:

- Intake: Also known as induction or suction. This stroke of the piston begins at top dead center (T.D.C.) and ends at bottom dead center (B.D.C.). In this stroke the intake valve must be in the open position while the piston pulls an air-fuel mixture into the cylinder by producing vacuum pressure into the cylinder through its downward motion. The piston is moving down as air is being sucked in by the downward motion against the piston.

- Compression: This stroke begins at B.D.C, or just at the end of the suction stroke, and ends at T.D.C. In this stroke the piston compresses the air-fuel mixture in preparation for ignition during the power stroke (below). Both the intake and exhaust valves are closed during this stage.

- Combustion: Also known as power or ignition. This is the start of the second revolution of the four stroke cycle. At this point the crankshaft has completed a full 360 degree revolution. While the piston is at T.D.C. (the end of the compression stroke) the compressed air-fuel mixture is ignited by a spark plug (in a gasoline engine) or by heat generated by high compression (diesel engines), forcefully returning the piston to B.D.C. This stroke produces mechanical work from the engine to turn the crankshaft.

- Exhaust: Also known as outlet. During the exhaust stroke, the piston, once again, returns from B.D.C. to T.D.C. while the exhaust valve is open. This action expels the spent air-fuel mixture through the exhaust valve.

These four strokes can be remembered by the colloquial phrase, "Suck, Squeeze, Bang, Blow".

Thermodynamic Analysis

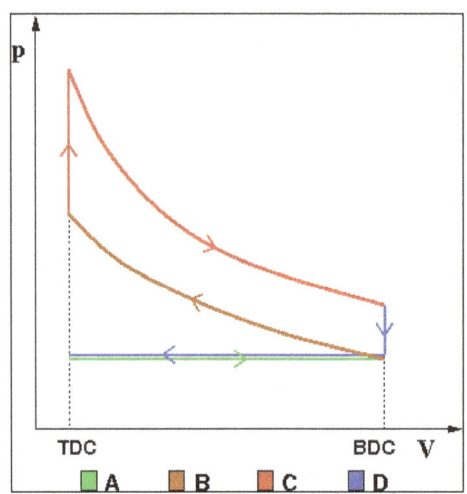

The idealized four-stroke Otto cycle p-V diagram: The intake (A) stroke is performed by an isobaric expansion, followed by the compression (B) stroke, performed by an adiabatic compression. Through the combustion of fuel an isochoric process is produced, followed by an adiabatic expansion, characterizing the power (C) stroke. The cycle is closed by an isochoric process and an isobaric compression, characterizing the exhaust (D) stroke.

The thermodynamic analysis of the actual four-stroke and two-stroke cycles is not a simple task. However, the analysis can be simplified significantly if air standard assumptions are utilized. The resulting cycle, which closely resembles the actual operating conditions, is the Otto cycle.

During normal operation of the engine, as the air/fuel mixture is being compressed, an electric spark is created to ignite the mixture. At low rpm this occurs close to TDC (Top Dead Centre). As engine rpm rises, the speed of the flame front does not change so the spark point is advanced earlier in the cycle to allow a greater proportion of the cycle for the charge to combust before the power stroke commences. This advantage is reflected in the various Otto engine designs; the atmospheric (non-compression) engine operates at 12% efficiency whereas the compressed-charge engine has an operating efficiency around 30%.

Fuel Considerations

A problem with compressed charge engines is that the temperature rise of the compressed charge can cause pre-ignition. If this occurs at the wrong time and is too energetic, it can damage the engine. Different fractions of petroleum have widely varying flash points (the temperatures at which the fuel may self-ignite). This must be taken into account in engine and fuel design.

The tendency for the compressed fuel mixture to ignite early is limited by the chemical composition of the fuel. There are several grades of fuel to accommodate differing performance levels of engines. The fuel is altered to change its self ignition temperature. There are several ways to do this. As engines are designed with higher compression ratios the result is that pre-ignition is much more likely to occur since the fuel mixture is compressed to a higher temperature prior to deliberate ignition. The higher temperature more effectively evaporates fuels such as gasoline, which increases the efficiency of the compression engine. Higher Compression ratios also means that the distance that the piston can push to produce power is greater (which is called the Expansion ratio).

The octane rating of a given fuel is a measure of the fuel's resistance to self-ignition. A fuel with a higher numerical octane rating allows for a higher compression ratio, which extracts more energy from the fuel and more effectively converts that energy into useful work while at the same time preventing engine damage from pre-ignition. High Octane fuel is also more expensive.

Diesel engines by their nature do not have concerns with pre-ignition. They have a concern with whether or not combustion can be started. The description of how likely Diesel fuel is to ignite is called the Cetane rating. Because Diesel fuels are of low volatility, they can be very hard to start when cold. Various techniques are used to start a cold Diesel engine, the most common being the use of a glow plug.

Design and Engineering Principles

Power Output Limitations

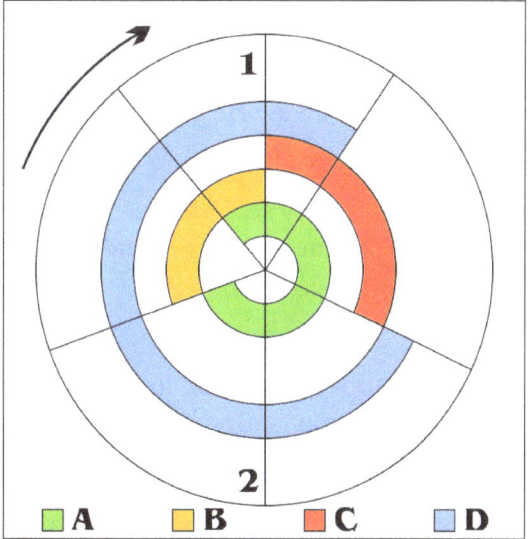

The four-stroke cycle 1=TDC 2=BDC A: Intake B: Compression C: Power D: Exhaust.

The maximum amount of power generated by an engine is determined by the maximum amount of air ingested. The amount of power generated by a piston engine is related to its size (cylinder volume), whether it is a two-stroke engine or four-stroke design, volumetric efficiency, losses, air-to-fuel ratio, the calorific value of the fuel, oxygen content of the air and speed (RPM). The speed is ultimately limited by material strength and lubrication. Valves, pistons and connecting rods suffer severe acceleration forces. At high engine speed, physical breakage and piston ring flutter can occur, resulting in power loss or even engine destruction. Piston ring flutter occurs when the rings oscillate vertically within the piston grooves they reside in. Ring flutter compromises the seal between the ring and the cylinder wall, which causes a loss of cylinder pressure and power. If an engine spins too quickly, valve springs cannot act quickly enough to close the valves. This is commonly referred to as 'valve float', and it can result in piston to valve contact, severely damaging the engine. At high speeds the lubrication of piston cylinder wall interface tends to break down. This limits the piston speed for industrial engines to about 10 m/s.

Intake/Exhaust Port Flow

The output power of an engine is dependent on the ability of intake (air–fuel mixture) and exhaust matter to move quickly through valve ports, typically located in the cylinder head. To increase an engine's output power, irregularities in the intake and exhaust paths, such as casting flaws, can be removed, and, with the aid of an air flow bench, the radii of valve port turns and valve seat configuration can be modified to reduce resistance. This process is called porting, and it can be done by hand or with a CNC machine.

Waste Heat Recovery of an Internal Combustion Engine

An internal combustion engine is on average capable of converting only 40-45% of supplied energy into mechanical work. A large part of the waste energy is in the form of heat that is released to the environment through coolant, fins etc. If we could somehow recover the waste heat we can improve the engine's performance. It has been found that even if 6% of the entirely wasted heat is recovered it can increase the engine efficiency greatly.

Many methods have been devised in order to extract waste heat out of an engine exhaust and use it further to extract some useful work, decreasing the exhaust pollutants at the same time. Use of the Rankine Cycle, turbocharging and thermoelectric generation can be very useful as a waste heat recovery system.

Though these systems are used more frequently some issues, like their low efficiency at lower heat supply rates and high pumping losses, remain a cause of concern.

Supercharging

One way to increase engine power is to force more air into the cylinder so that more power can be produced from each power stroke. This can be done using some type of air compression device known as a supercharger, which can be powered by the engine crankshaft.

Supercharging increases the power output limits of an internal combustion engine relative to its displacement. Most commonly, the supercharger is always running, but there have been designs that allow it to be cut out or run at varying speeds (relative to engine speed). Mechanically driven supercharging has the disadvantage that some of the output power is used to drive the supercharger, while power is wasted in the high pressure exhaust, as the air has been compressed twice and then gains more potential volume in the combustion but it is only expanded in one stage.

Turbocharging

A turbocharger is a supercharger that is driven by the engine's exhaust gases, by means of a turbine.

A turbocharger is incorporated into the exhaust system of a vehicle to make use of the expelled exhaust. It consists of a two piece, high-speed turbine assembly with one side that compresses the intake air, and the other side that is powered by the exhaust gas outflow.

When idling, and at low-to-moderate speeds, the turbine produces little power from the small exhaust volume, the turbocharger has little effect and the engine operates nearly in a naturally aspirated manner. When much more power output is required, the engine speed and throttle opening are increased until the exhaust gases are sufficient to 'spool up' the turbocharger's turbine to start compressing much more air than normal into the intake manifold. Thus, additional power (and speed) is expelled through the function of this turbine.

Turbocharging allows for more efficient engine operation because it is driven by exhaust pressure that would otherwise be (mostly) wasted, but there is a design limitation known as turbo lag. The increased engine power is not immediately available due to the need to sharply increase engine RPM, to build up pressure and to spin up the turbo, before the turbo starts to do any useful air compression. The increased intake volume causes increased exhaust and spins the turbo faster, and so forth until steady high power operation is reached. Another difficulty is that the higher exhaust pressure causes the exhaust gas to transfer more of its heat to the mechanical parts of the engine.

Rod and Piston-to-stroke Ratio

The rod-to-stroke ratio is the ratio of the length of the connecting rod to the length of the piston stroke. A longer rod reduces sidewise pressure of the piston on the cylinder wall and the stress forces, increasing engine life. It also increases the cost and engine height and weight.

A "square engine" is an engine with a bore diameter equal to its stroke length. An engine where the bore diameter is larger than its stroke length is an oversquare engine, conversely, an engine with a bore diameter that is smaller than its stroke length is an undersquare engine.

Valve Train

The valves are typically operated by a camshaft rotating at half the speed of the crankshaft. It has a series of cams along its length, each designed to open a valve during the appropriate part of an intake or exhaust stroke. A tappet between valve and cam is a contact surface on which the cam slides to open the valve. Many engines use one or more camshafts "above" a row (or each row) of cylinders, as in the illustration, in which each cam directly actuates a valve through a flat tappet. In other engine designs the camshaft is in the crankcase, in which case each cam usually contacts a push rod, which contacts a rocker arm that opens a valve, or in case of a flathead engine a push rod is not necessary. The overhead cam design typically allows higher engine speeds because it provides the most direct path between cam and valve.

Valve Clearance

Valve clearance refers to the small gap between a valve lifter and a valve stem that ensures that the valve completely closes. On engines with mechanical valve adjustment, excessive clearance causes noise from the valve train. A too small valve clearance can result in the valves not closing properly, this results in a loss of performance and possibly overheating of exhaust valves. Typically, the clearance must be readjusted each 20,000 miles (32,000 km) with a feeler gauge.

Most modern production engines use hydraulic lifters to automatically compensate for valve train component wear. Dirty engine oil may cause lifter failure.

Energy Balance

Otto engines are about 30% efficient; in other words, 30% of the energy generated by combustion is converted into useful rotational energy at the output shaft of the engine, while the remainder being losses due to waste heat, friction and engine accessories. There are a number of ways to recover some of the energy lost to waste heat. The use of a Turbocharger in Diesel engines is very effective by boosting incoming air pressure and in effect, provides the same increase in performance as having more displacement. The Mack Truck company, decades ago, developed a turbine system that converted waste heat into kinetic energy that it fed back into the engine's transmission. In 2005, BMW announced the development of the turbosteamer, a two-stage heat-recovery system similar to the Mack system that recovers 80% of the energy in the exhaust gas and raises the efficiency of an Otto engine by 15%. By contrast, a six-stroke engine may reduce fuel consumption by as much as 40%.

Modern engines are often intentionally built to be slightly less efficient than they could otherwise be. This is necessary for emission controls such as exhaust gas recirculation and catalytic converters that reduce smog and other atmospheric pollutants. Reductions in efficiency may be counteracted with an engine control unit using lean burn techniques.

In the United States, the Corporate Average Fuel Economy mandates that vehicles must achieve an average of 34.9 mpg$_{-US}$ (6.7 L/100 km; 41.9 mpg$_{-imp}$) compared to the current standard of 25 mpg$_{-US}$ (9.4 L/100 km; 30.0 mpg$_{-imp}$). As automakers look to meet these standards by 2016, new ways of engineering the traditional internal combustion engine (ICE) have to be considered. Some potential solutions to increase fuel efficiency to meet new mandates include firing after the piston is farthest from the crankshaft, known as top dead centre, and applying the Miller cycle. Together, this redesign could significantly reduce fuel consumption and NO x emissions.

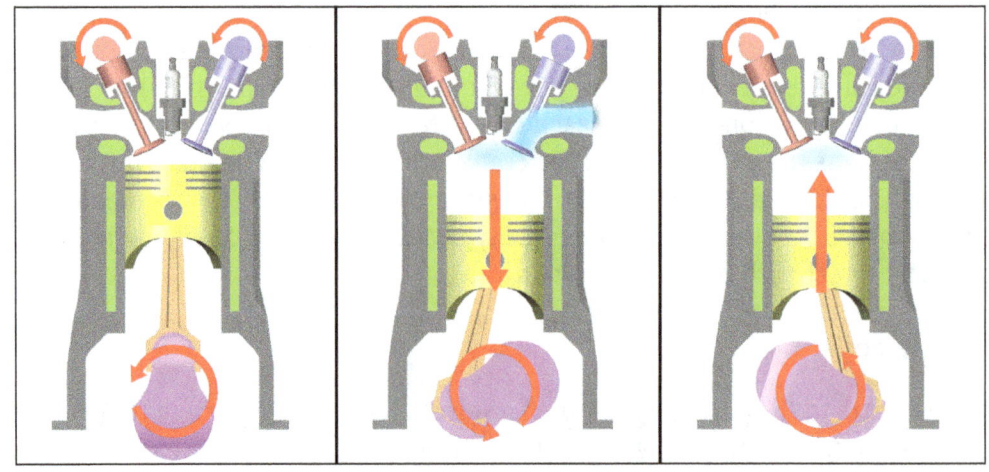

Starting position, intake stroke, and compression stroke.

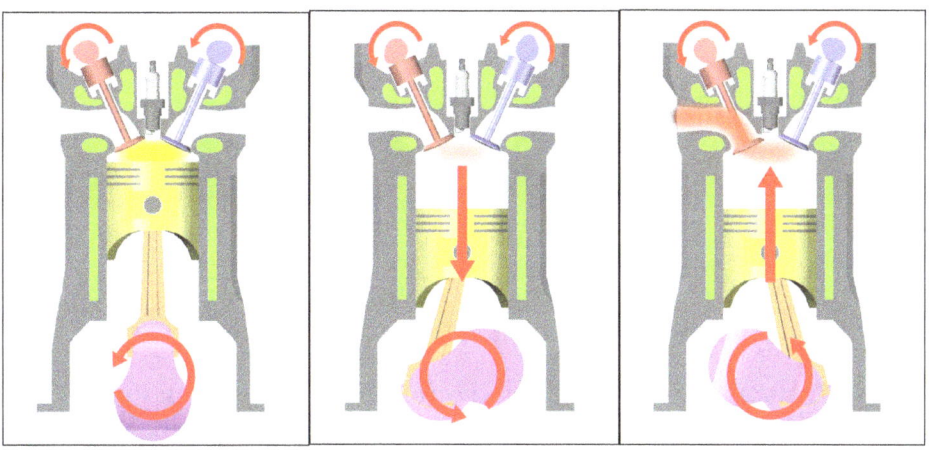

Ignition of fuel, power stroke, and exhaust stroke.

H ENGINE

An H engine (or H-block) is an engine configuration in which the cylinders are aligned so that if viewed from the front, they appear to be in a vertical or horizontal letter H.

An H engine can be viewed as two flat engines, one atop or beside the other. The "two engines" each have their own crankshaft, which are then geared together at one end for power-take-off. The H configuration allows the building of multi-cylinder engines that are shorter than the alternatives, sometimes delivering advantages on aircraft. For race-car applications there is the disadvantage of a higher centre of gravity, not only because one crankshaft is located atop the other, but also because the engine must be high enough off the ground to allow clearance underneath for a row of exhaust pipes. The power-to-weight ratio is not as good as simpler configurations employing one crankshaft. There is excellent mechanical balance, especially desirable and otherwise difficult to achieve in a four-cylinder engine.

Two straight engines can be similarly joined to provide a U engine.

List of H engines

Other Engines

Brough Superior H-4 motorcycle engine.

A BRM H16 engine, mounted in the back of a BRM P83 Formula One car.

- The British Racing Motors (BRM) H-16 Formula One engine won the 1966 US Grand Prix with Jim Clark in a Lotus 43. As a racing-car engine it was hampered by a high center of gravity, and it was heavy and complex, with gear-driven twin overhead cams for each of four cylinder heads, two gear-coupled crankshafts, and mechanical fuel injection.

- The Brough Superior Golden Dream motorcycle, first shown in 1938. A 1,000 cc H-4 design and a few units were produced in early 1939. Any development planned was interrupted by World War II and subsequent years of austerity.

- Wooler built a motorcycle prototype with a similar configuration to the Brough Superior Golden Dream and exhibited it at the British International Motor Show at Earls Court Exhibition Centre in 1948 and again in 1951. This was replaced by a flat-four engined prototype at the 1953 show.

- German firm Konig, who specialised in racing outboard motors, built a few 1000cc H-8s c. 1970s, which were basically two of their VC500 flat fours mounted one above the other, with the direction of rotation reversed on one of them. Each half of the engine was a water cooled 2-stroke with rotating disc valve driven by a toothed belt via two 45/90 degree pulleys, plus two siamesed expansion chamber exhausts, fed by two single choke carbs. Both cylinders at each end of each engine fired at the same time, hence the siamesed exhausts for each pair. At least one H-8 found its way into a motorcycle & sidecar racing combo. The VC500 was rated initially at 68 bhp, later 75 & c.100 in the single seater Voigt Renwick Special designed for Hillclimbing, so the H-8 might achieve double these figures, although with very narrow power band and some reliability issues.

PUSHROD ENGINE

An overhead valve engine (OHV engine), or pushrod engine, is a reciprocating piston engine whose poppet valves are situated in the cylinder head. An OHV engine's valvetrain operates its valves via a camshaft within the cylinder block, cam followers (or "tappets"), pushrods, and rocker arms.

The OHV engine was an advance over the older flathead engine, whose valves were situated in the cylinder block. Some early "OHV" engines known as "F-heads" used both side-valves and overhead valves. A variation over the OHV design is the overhead camshaft, or "OHC", engine, whose camshaft lies in the cylinder head itself, above the valves. To avoid confusion, OHC engines are not referred to as OHV despite also having their valves in the head.

Advantages

OHV engines have some advantages over OHC engines:

- Smaller overall packaging: Because of the camshaft's location inside the engine block, OHV engines are more compact than an overhead cam engine of comparable displacement. For example, Ford's 4.6 L OHC modular V8 is larger than the 5.0 L I-head Windsor V8 it replaced. GM's 4.6 L OHC Northstar V8 is slightly taller and wider than GM's larger displacement 5.7 to 7.0 L I-head LS V8. The Ford Ka uses the Kent Crossflow/Endura-E OHV engine to fit under its low bonnet line. Because of the generally more compact size of an engine of a given displacement, a pushrod engine of given external dimensions can have significantly greater displacement than an OHC engine of the same external size. As a result, the pushrod engine can sometimes produce just as much power as the OHC engine, but with greater torque (contrary to popular belief, this is simply due to the greater displacement of the pushrod engine versus the OHC engine rather than any inherent advantage of the pushrod design for torque production).

- Only a single head casting is needed on vee and flat pushrod engines: Because of the need to drive both cams on one side of the engine, the camshaft orientation on OHC heads must be the same for both heads of a flat and vee engine. This means that the heads must be (more or less) mirror images of each other. And this requires that two different head castings be produced. Pushrod heads can simply be flipped around, which allows a single casting to be used on both cylinder banks.

- Simpler drive system: OHV engines have a less complex drive system for the camshaft when compared with OHC engines. Most OHC engines drive the camshaft or camshafts using a timing belt, a chain, or multiple chains. These systems require the use of tensioners, which add complexity. In contrast, an OHV engine has the camshaft positioned close to the crankshaft, which may be driven by a much shorter chain or even direct gear connection. However, this is somewhat negated by a more complex valvetrain requiring pushrods.

- Hydraulic lifters: Although RPM capability is limited by the use of hydraulic lifters, the valve lash is self-adjusting for the lifetime of the engine, reducing a significant maintenance requirement. Some OHC engines also use hydraulic lifters/lash adjusters, but the implementation is more complex in OHC designs.

- Simpler lubrication system: Because OHV engines have no camshafts in the heads, the heads have much more modest lubrication requirements than the heads in OHC engines. Therefore, there is no need for oil galleys to supply the heads with oil or oil galleys in the head to provide lubrication for the cam bearings. OHV heads only need lubrication for the rocker arms at the pushrod end, trunnion, and rocker tip. This lubrication to is typically provided through the pushrods themselves rather than a dedicated lubrication system in the head. And lubrication for the camshaft is provided through the same block galleys that provide oil for the main bearings. The more modest lubrication needs of an OHV engine also mean that a smaller, lower-capacity oil pump can be used.

Limitations

Some specific problems that remain with overhead valve (OHV) engines:

- Limited engine speeds or RPM: OHV engines have more valvetrain moving parts. OHV engines also typically use only a single intake and exhaust valve, which results in large (and heavy) valves, valve springs, and retainers. Thus, the valvetrain in an OHV engine has greater inertia and mass. As a result, they suffer more easily from valve "float", and may exhibit a tendency for the pushrods, if improperly designed, to flex or snap at high engine speeds. Therefore, OHV engine designs cannot spin at engine speeds as high as OHC Modern OHV engines are usually limited to about 6,000 to 8,000 revolutions per minute (rpm) in production cars, and 9,000 rpm to 10,500 rpm in racing applications. In contrast, many modern DOHC engines may have rev limits from 6,000 rpm to 9,000 rpm in road car engines, and in excess of 20,000 rpm (though now limited to 15,000 rpm) in current Formula One engines using pneumatic valve springs. High-revving pushrod engines are normally solid (mechanical) lifter designs, flat and roller. In 1969, Chevrolet offered a Corvette and a Camaro model with a solid lifter cam pushrod V8 (the ZL-1) that could rev to 8,000 rpm. The Volvo B18 and B20 engines can rev to more than 7,000 rpm with their solid lifter camshaft. However, the LS7 of the C6 Corvette Z06 is the first production hydraulic roller cam pushrod engine to have a redline of 7,100 rpm. The Honda CX500 motorcycle engine has a 9650rpm redline, well above the usual limits for auto engines, due to the lighter weight of components.

- Limited cylinder head design flexibility: overhead camshaft (OHC) engines benefit substantially from the ability to use multiple valves per cylinder, as well as much greater freedom of component placement, and intake and exhaust port geometry. Most modern OHV engines have two valves per cylinder, while many OHC engines can have three, four or even five valves per cylinder to achieve greater power. Though multi-valve OHV engines exist, their use is somewhat limited due to their complexity and is mostly restricted to low- and medium-speed diesel engines, with a few notable exceptions such as the four-valve-per-cylinder Honda CX500 motorcycle, and the Harley-Davidson Milwaukee-Eight engine. In OHV engines, the size and shape of the intake ports as well as the position of the valves are limited by the pushrods and the need to accommodate them in the head casting. Spark plug placement is also less ideal in pushrod engines. This is important, since a centrally located spark plug improves combustion efficiency and reduces both emissions and tendency to detonate by reducing flame travel distance (which also reduces combustion time). DOHC engines with four valves per cylinder can have a truly centrally located spark plug because this space is free from both valves in the combustion chamber and valvetrain above this central area. Even SOHC engines with four valves per cylinder can usually accommodate a central spark plug. But since pushrod engines almost always have only two valves per cylinder, it is impossible to have a central spark plug.

- Noise and refinement: OHV engines are generally noisier than their OHC counterparts owing to the increased complexity of the valvetrain and the adoption of chain- or gear-based camshaft drive.

- Maintenance: The location of the camshaft in the cylinder block often necessitates removal of the engine whenever camshaft work is required. This is particularly true for front wheel drive applications with a transversely mounted engine. Longitudinally mounted OHV engines suffer less from this problem as the camshaft can be withdrawn from the front of the engine after removal of the radiator. Additionally, replacement of lifters generally requires removal of the cylinder heads. And cam bearing replacement generally requires the removal and complete teardown of the engine.

- Increased valvetrain friction and wear: Because OHV engines generally use only a single intake and exhaust valve, the valves are larger and heavier than those used in multivalve OHC engines. Furthermore, when the valve is closed, the spring must accelerate not only the valve and rocker arm, but also the pushrod and lifter. Therefore, OHV engines must use heavier valve springs than OHC engines (multivalve or otherwise). As a result, valvetrain friction and wear is increased. This is especially true with high-performance OHV engines utilizing high lift/duration/ramp rate cams with heavier than stock valve springs.

- Limited ability to use variable valve timing: Because OHV engines use a single camshaft for all valves, the ability to independently vary intake and exhaust valve timing is limited, although it is possible to vary the phase of one set of valves while the other set's timing stays constant (as in the "CamInCam" or "DuoCam" system). Also, because all cam lobes are on a single camshaft, there is little to no room for the extra high RPM lobes required for two stage systems like Hondas's VTEC. Because of these factors, variable valve timing on OHV engines is limited to cam phasing systems that change intake and exhaust valve timing with no ability to vary lift or duration.

RECIPROCATING ENGINE

A reciprocating engine, also often known as a piston engine, is typically a heat engine (although there are also pneumatic and hydraulic reciprocating engines) that uses one or more reciprocating pistons to convert pressure into a rotating motion. The main types are: the internal combustion engine, used extensively in motor vehicles; the steam engine, the mainstay of the Industrial Revolution; and the niche application Stirling engine. Internal combustion engines are further classified in two ways: either a spark-ignition (SI) engine, where the spark plug initiates the combustion; or a compression-ignition (CI) engine, where the air within the cylinder is compressed, thus heating it, so that the heated air ignites fuel that is injected then or earlier.

Common Features in all Types

There may be one or more pistons. Each piston is inside a cylinder, into which a gas is introduced, either already under pressure (e.g. steam engine), or heated inside the cylinder either by ignition of a fuel air mixture (internal combustion engine) or by contact with a hot heat exchanger in the cylinder (Stirling engine). The hot gases expand, pushing the piston to the bottom of the cylinder. This position is also known as the Bottom Dead Center (BDC), or where the piston forms the largest volume in the cylinder. The piston is returned to the cylinder top (Top Dead Centre) (TDC) by a flywheel, the power from other pistons connected to the same shaft or (in a double acting cylinder) by the same process acting on the other side of the piston. This is where the piston forms the smallest volume in the cylinder. In most types the expanded or "exhausted" gases are removed from the cylinder by this stroke. The exception is the Stirling engine, which repeatedly heats and cools the same sealed quantity of gas. The stroke is simply the distance between the TDC and the BDC, or the greatest distance that the piston can travel in one direction.

In some designs the piston may be powered in both directions in the cylinder, in which case it is said to be double-acting.

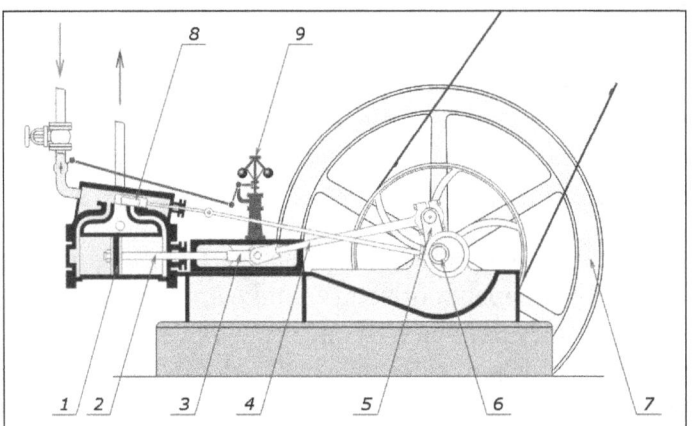

Steam piston engine.

A labeled schematic diagram of a typical single-cylinder, simple expansion, double-acting high pressure steam engine. Power takeoff from the engine is by way of a belt.

- Piston.

- Piston rod.

- Crosshead bearing.

- Connecting rod.

- Crank.

- Eccentric valve motion.

- Flywheel.

- Sliding valve.

- Centrifugal governor.

In most types, the linear movement of the piston is converted to a rotating movement via a connecting rod and a crankshaft or by a swashplate or other suitable mechanism. A flywheel is often used to ensure smooth rotation or to store energy to carry the engine through an un-powered part of the cycle. The more cylinders a reciprocating engine has, generally, the more vibration-free (smoothly) it can operate. The power of a reciprocating engine is proportional to the volume of the combined pistons' displacement.

A seal must be made between the sliding piston and the walls of the cylinder so that the high pressure gas above the piston does not leak past it and reduce the efficiency of the engine. This seal is usually provided by one or more piston rings. These are rings made of a hard metal, and are sprung into a circular groove in the piston head. The rings fit closely in the groove and press lightly against the cylinder wall to form a seal, and more heavily when higher combustion pressure moves around to their inner surfaces.

It is common to classify such engines by the number and alignment of cylinders and total volume of displacement of gas by the pistons moving in the cylinders usually measured in cubic centimetres (cm³ or cc) or litres (l) or (L) (US: liter). For example, for internal combustion engines, single and two-cylinder designs are common in smaller vehicles such as motorcycles, while automobiles typically have between four and eight, and locomotives, and ships may have a dozen cylinders or more. Cylinder capacities may range from 10 cm³ or less in model engines up to thousands of liters in ships' engines.

The compression ratio affects the performance in most types of reciprocating engine. It is the ratio between the volume of the cylinder, when the piston is at the bottom of its stroke, and the volume when the piston is at the top of its stroke.

The bore/stroke ratio is the ratio of the diameter of the piston, or "bore", to the length of travel within the cylinder, or "stroke". If this is around 1 the engine is said to be "square", if it is greater than 1, i.e. the bore is larger than the stroke, it is "oversquare". If it is less than 1, i.e. the stroke is larger than the bore, it is "undersquare".

Cylinders may be aligned in line, in a V configuration, horizontally opposite each other, or radially around the crankshaft. Opposed-piston engines put two pistons working at opposite ends of the same cylinder and this has been extended into triangular arrangements such as the Napier Deltic. Some designs have set the cylinders in motion around the shaft, such as the Rotary engine.

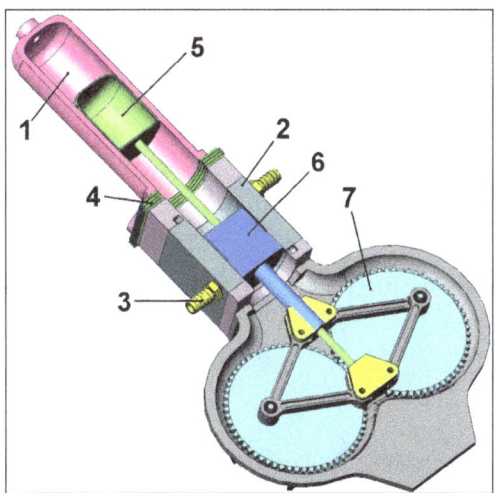

Stirling piston engine Rhombic Drive – Beta Stirling Engine Design, showing the second displacer piston (green) within the cylinder, which shunts the working gas between the hot and cold ends, but produces no power itself.

- Hot cylinder wall.

- Cold cylinder wall.

- Displacer piston.

- Power piston.

- Flywheels.

In steam engines and internal combustion engines, valves are required to allow the entry and exit of gases at the correct times in the piston's cycle. These are worked by cams, eccentrics or cranks driven by the shaft of the engine. Early designs used the D slide valve but this has been largely superseded by Piston valve or Poppet valve designs. In steam engines the point in the piston cycle at which the steam inlet valve closes is called the cutoff and this can often be controlled to adjust the torque supplied by the engine and improve efficiency. In some steam engines, the action of the valves can be replaced by an oscillating cylinder.

Internal combustion engines operate through a sequence of strokes that admit and remove gases to and from the cylinder. These operations are repeated cyclically and an engine is said to be 2-stroke, 4-stroke or 6-stroke depending on the number of strokes it takes to complete a cycle.

In some steam engines, the cylinders may be of varying size with the smallest bore cylinder working the highest pressure steam. This is then fed through one or more, increasingly larger bore cylinders successively, to extract power from the steam at increasingly lower pressures. These engines are called Compound engines.

Aside from looking at the power that the engine can produce, the Mean Effective Pressure (MEP), can also be used in comparing the power output and performance of reciprocating engines of the same size. The mean effective pressure is the fictitious pressure which would produce the same amount of net work that was produced during the power stroke cycle. This is shown by:

$$W_{net} = MEP \times \text{Piston Area} \times \text{Stroke} = MEP \times \text{Displacement Volume}$$

and therefore:

$$MEP = W_{net}/\text{Displacement Volume}.$$

Whichever engine with the larger value of MEP produces more net work per cycle and performs more efficiently.

Engine Capacity

For piston engines, an engine's capacity is the engine displacement, in other words the volume swept by all the pistons of an engine in a single movement. It is generally measured in litres (l) or cubic inches (c.i.d., cu in, or in^3) for larger engines, and cubic centimetres (abbreviated cc) for smaller engines. All else being equal, engines with greater capacities are more powerful and consumption of fuel increases accordingly (although this is not true of every Reciprocating engine), although power and fuel consumption are affected by many factors outside of engine displacement.

Power

Reciprocating engines can be characterized by their specific power, which is typically given in kilowatts per litre of engine displacement (in the U.S. also horsepower per cubic inch). The result offers an approximation of the peak power output of an engine. the fuel efficiency is high often requires a lean fuel-air ratio, and thus lower power density. A modern high-performance car engine makes in excess of 75 kW/L (1.65 hp/in^3).

SINGLE CYLINDER ENGINE

A single-cylinder engine is a piston engine with one cylinder. It is often used on motorcycles, scooters, go-karts, ATVs, radio-controlled vehicles, portable tools and garden machinery (such as lawnmowers, rototillers and string trimmers).

Characteristics

Villiers engine in a 1959 Bond Minicar.

Compared with multi-cylinder engines, single-cylinder engines are usually more simple and compact. Due to the greater potential for airflow around all sides of the cylinder, air cooling is often more effective for single cylinder engines than multi-cylinder engines. This reduces the weight and complexity of air-cooled single-cylinder engines, compared with liquid-cooled engines.

Drawbacks of single-cylinder engines include a more pulsating power delivery through each cycle and higher levels of vibration. The uneven power delivery means that often a single-cylinder engine requires a heaver flywheel than a comparable multi-cylinder engine, resulting in relatively slower changes in engine speed. To reduce the vibration level, they often make greater use of balance shafts than multi-cylinder engines, as well as more extreme methods such as a dummy connecting rod (for example the Ducati Supermono). These balancing devices can reduce the benefits of single-cylinder engines regarding lower weight and complexity.

Most single-cylinder engines used in motor vehicles are fueled by petrol (and use a four-stroke cycle), however diesel single-cylinder engines are also used in stationary applications (such as the Lombardini 3LD and 15LD).

A variation known as the split-single makes use of two pistons which share a single combustion chamber.

Uses

Yamaha SRX600 motorcycle engine.

Early motorcycles, automobiles and other applications such as marine engines all tended to be single-cylinder. The configuration is almost exclusively used in portable tools, along with garden machinery such as lawn mowers. Single cylinder engines also remain in widespread use in motorcycles, motor scooters, go-karts, auto rickshaws, and radio-controlled models. From 1921-1960, the Lanz Bulldog tractor used a large horizontally-mounted single cylinder two-stroke engine. However they are rarely used in automobiles and tractors these days, due to developments in engine technology.

Single cylinder engines remain the most common engine layout in motor scooters and low-powered motorcycles. The Honda Super Cub (the motor vehicle with the highest overall sales since its introduction in 1958) uses a 49 cc (3.0 cu in) four-stroke single-cylinder engine. There are also several single-cylinder sportbikes (such as the KTM 690 Duke R), dual-sport motorcycles (such as the BMW G650GS) and the classic-styled Royal Enfield 500 Bullet.

STRAIGHT ENGINE

The straight or inline engine is an internal-combustion engine with all cylinders aligned in one row and having no offset. Usually found in four, six and eight cylinder configurations, they have been used in automobiles, locomotives and aircraft, although the term in-line has a broader meaning when applied to aircraft engines.

A straight engine is considerably easier to build than an otherwise equivalent horizontally opposed or V engine, because both the cylinder bank and crankshaft can be milled from a single metal casting, and it requires fewer cylinder heads and camshafts. In-line engines are also smaller in overall physical dimensions than designs such as the radial, and can be mounted in any direction. Straight configurations are simpler than their V-shaped counterparts. Although six-cylinder engines are inherently balanced,the four-cylinder models are inherently off balance and rough, unlike 90-degree V fours and horizontally opposed 'boxer' four cylinders.

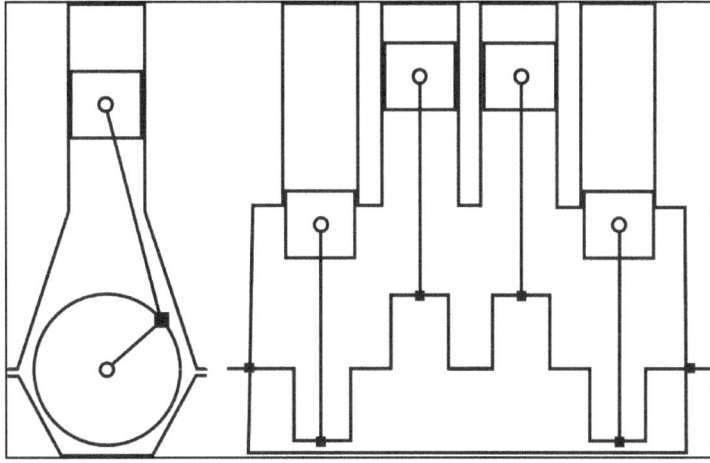

4-cylinder straight engine scheme.

Automobile Use

The inline-four engine is the most common four-cylinder configuration, whereas the straight-6 has largely given way to the V6 engine, which although not as naturally smooth-running is smaller in both length and height and easier to fit into the engine bay of smaller modern cars. Some manufacturers, including Acura, Audi, Ford, Mercedes-Benz, Volkswagen and Volvo, have also used straight-five configurations. The General Motors Atlas family includes straight-four, straight-five, and straight-six engines. Some small cars have inline three engines.

Once, the straight-eight was the prestige engine arrangement; it could be made more cheaply than a V-engine by luxury car makers, who would focus on other specifics than the geometric ones, and even built engines more powerful than any V8 engine. In the 1930s, Duesenberg used a cylinder block made from aluminium alloy, with four valves per cylinder and hemispherical heads to produce the most powerful engine on the market. It was thus a selling point for Pontiac to introduce the cheapest straight-eight in 1933. However, following World War II, the straight-eight was supplanted by the lighter and more compact V8 engine, which allowed shorter engine bays to be used in the design.

When a straight engine is mounted at an angle from the vertical it is called a slant engine. Chrysler's Slant 6 was used in many models in the 1960s and 1970s. Honda also often mounts its straight-four and straight-five engines at a slant, as on the Honda S2000 and Acura Vigor. SAAB initially used the Triumph Slant-4 engine tilted at 45 degrees for the Saab 99, but later versions of the engine were less tilted.

Two main factors have led to the recent decline of the straight-six in automotive applications. First, Lanchester balance shafts, an old idea reintroduced by Mitsubishi in the 1980s to overcome the natural imbalance of the inline-four engine and rapidly adopted by many other manufacturers, have made both inline-four and V6 engines smoother-running; the greater smoothness of the straight-six layout is no longer as great an advantage. Second, fuel consumption became more important, as cars became smaller and more space-efficient. The engine bay of a modern small or medium car, typically designed for an inline-four, often does not have room for a straight-six, but can fit a V6 with only minor modifications.

Some manufacturers (originally Lancia, and more recently Volkswagen with the VR6 engine) have attempted to combine advantages of the straight and V configurations by producing a narrow-angle V; this is more compact than either configuration, but is less smooth (without balancing) than either.

Straight-6 engines are used in some models from BMW, Ford, Jeep, Chevrolet, GMC, Toyota, Suzuki and Volvo Cars.

STRAIGHT-SIX ENGINE

The straight-six engine or inline-six engine (often abbreviated I6 or L6) is an internal combustion engine with the cylinders mounted in a straight line along the crankcase with all the pistons driving a common crankshaft (straight engine).

The bank of cylinders may be oriented at any angle, and where the bank is inclined away from the vertical, the engine is sometimes called a slant-six (although this is also a Chrysler-specific design). The straight-six layout is the simplest engine layout that possesses both primary and secondary mechanical engine balance, resulting in much less vibration than engines with fewer cylinders.

A BMW M20B25 engine with the cylinder head removed, showing the pistons in the six cylinders of the engine.

Displacement Range

In automobiles, the straight-six design is used for engine displacements ranging from approximately 2 to 5 litres (120 to 310 cu in). It is also sometimes used for smaller engines but these, although very smooth running, tend to be rather expensive to manufacture in terms of cost-to-power ratio. Since the length of an engine is roughly proportional to the number of cylinders in one bank, the straight-six is necessarily longer than alternative layouts such as I4, V4, V6, or V8.

One of the smallest production straight-sixes was found in the Benelli 750 Sei motorcycle, displacing 747.7 cc (45.63 cu in) (0.7477 L). Honda and Mike Hailwood raced in the 1960s with the RC166 250 cc (15 cu in) (0.25 L) six-cylinder, 24-valve motorcycle engine. Pre-World War II engines could be quite large by modern standards — such as the Rolls Royce Silver Ghost's 7.4 L engine and the 824 cu in (13.5 L) of the 1910s Peerless, Pierce, and Fageol.

The largest modern passenger-car straight-sixes include the 4.6 L (282 cu in) VAM version of AMC's engines, the 4.2 litre powerplants found in several Jaguars and AMCs, 4.0 TVR Speed Six, 4.0 Ford Barra, Chevrolet 250, Chevrolet Vortec 4200, 4.3 Chrysler Hemi Six, 4.2 Toyota Land Cruiser (both diesel and petrol), 4.5, and 4.8 Nissan, 4.8 Chevrolet, 4.9 Ford, and the 5.0 L of Hudson H-145. As of 2009, the Cummins B Series engine used in Dodge Ram pickup trucks displaced up to 6.7 L.

The Gipsy Six and Gipsy Queen, made by the de Havilland Engine Company from 1935 until 1950, were inverted straight-six engines displacing 560.6 cubic inches (9.187 L). They were used in a variety of aircraft including the de Havilland Dragon Rapide and the Cierva W.9 experimental helicopter. The standard straight-six configuration World War I aviation engines used by the German Empire's Luftstreitkräfte aircraft possessed even larger displacements, with the most-used Mercedes D.III family of liquid-cooled, dual ignition, SOHC-valvetrain inline-six engines having a massive 14.8 litres (903 cu in).

Because it is a fully balanced configuration, the straight-six can be scaled up to very large sizes for heavy truck, industrial and marine use, such as the 16 L (980 cu in) Volvo diesel engine and the 15 L Cummins ISX used in heavy vehicles. The largest are used to power ships, and use fuel oil. The straight-six can also be viewed as a scalable modular component of larger motors which stack several straight-sixes together, e.g. flat- or V-12s, W-18s, etc.

Cars

Straight-six engines were introduced much earlier than V6 engines. While the first straight-six was manufactured in 1903 by Spyker, it was not until 1950 that a production V6 was introduced. V6s (unlike crossplane V8 engines) had intrinsic vibration problems not present in the straight-six.

The length of the straight-six was not a major concern in the older front-engine/rear-wheel drive vehicles, but the modern move to the more space-efficient front-engine/front-wheel drive and transverse engine (left-to-right versus front-to-back) configurations in smaller cars made the length of the V6 a major advantage. The overall length of an engine may be approximated by adding the bore pitch times the number of cylinders in one bank, plus the width of one connecting rod. As a result, in recent decades automobile manufacturers have replaced most of their straight-six engines (and many of their V8s) with V6 engines; Nissan have replaced their earlier turbocharged inline-6 models with larger displacement naturally aspirated V6 engines while maintaining the FR setup.

Exceptions to the shift to V engines include BMW, which specializes in high-performance straight-sixes used in a lineup of front-engine/rear-wheel-drive vehicles, almost all of BMW's current 6-cylinder model line-up use the straight configuration, Volvo, which designed a compact straight-six engine/transmission package to fit transversely in its larger cars, and the Australian Ford Falcon, which still uses a straight-six configuration. TVR used a straight-six configuration exclusively in their final cars before their demise.

Manufacturers began to replace V8 engines with straight-6 engines and V6 engines with straight-4 engines, while V8 engines became smaller. This was a part of a trend toward higher efficiency engines with fewer cylinders, but the same power output as previous larger engines as fuel economy standards became more stringent. A result of modular engine designs was that straight-6 engines could be built on the same assembly lines as straight-4 engines, while V8 engines smaller than previous V8s could be built with the same components as straight-4 engines in the same family. In a reversal of previous trends, Mercedes-Benz announced a return to inline-6 engines in October 2016.

Increasingly straight-six engines are also being replaced by turbocharged in-line four cylinder engines that offer comparable top-end power output and reduced low-end torque but with better fuel efficiency, due to smaller displacements and lower friction from the reduced number of cylinders. The poor secondary harmonic balance of four-cylinder engines is sometimes addressed with the use of balance shafts, but they are not inherently as smooth as an inline-6.

Balance and Smoothness

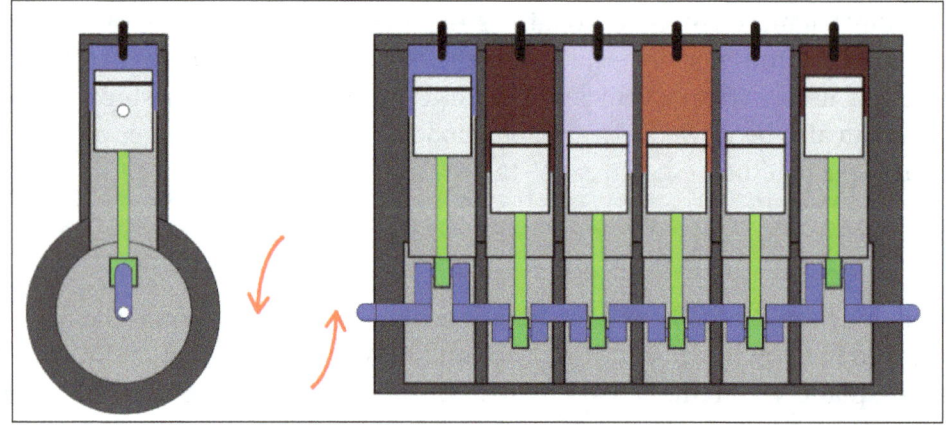

Inline 6 Cylinder with firing order 1-5-3-6-2-4.

An inline six engine is in practically perfect primary and secondary mechanical balance, without the use of a balance shaft. The engine is in primary couple balance because the front and rear trio of cylinders are mirror images, and the pistons move in pairs (but of course, 360° out of phase and on different strokes of the 4-stroke cycle). That is, piston #1 mirrors #6, #2 mirrors #5, and #3 mirrors #4, largely eliminating the polar rocking motion that would otherwise result.

Secondary imbalance is largely avoided because the crankshaft has six crank throws arranged in three planes offset at 120°. The result is that the bulk of the secondary forces that are caused by the pistons' deviation from purely sinusoidal motion sum to zero. Specifically, the second-order (twice crank speed) and fourth-order inertial free forces sum to zero, but the sixth-order and up are non-zero. This is typically a tiny contribution in most applications, but may be significant with very large displacements, despite the usual and advantageous use of long connecting rods reducing the secondary (second-order and up) oscillation in the piston motion in those applications.

A MAN B&W 6S60MC inline six-cylinder slow-speed diesel engine. This example is used on a 70,000 t (deadweight) bulk carrier, and produces 9,014.8 kW (12,089.0 hp) at 90.3 rpm (1.5 Hz) (703,130 ft.lbs torque).

An inline four cylinder, or even a V6 engine with a crank-speed balance shaft, will experience significant secondary dynamic imbalance, resulting in engine vibration. As a general rule, the forces arising from any dynamic imbalance increase as the square of the engine speed — for example, if the speed doubles, vibration will increase by a factor of four. In contrast, inline six engines have no primary or (significant) secondary imbalances, and with carefully designed crankshaft vibration dampers to absorb torsional vibration, will run more smoothly at the same crankshaft speed (rpm). This characteristic has made the straight-six popular in some European sports-luxury cars, where smooth high-speed performance is very desirable. As engine reciprocating forces increase with the cube of piston bore, the straight-six is a preferred configuration for large truck engines.

Inertial Torque

One other aspect that affects drivetrain smoothness is the variation in rotational speed arising from the sharing of piston kinetic energy amongst the different phases. There are only three piston phases in a four stroke inline six and, whilst the nature of the piston motion can never result in a 3rd-order imbalance (the pistons have no oscillation at that rate), it does appear in the kinetic energy exchange between the pistons and crankshaft (mathematically because of the squaring of velocity resulting in a specific intermodulation of frequencies). This means that a constant kinetic energy level for the rotating and reciprocating masses will result in the flywheel rotation speeding up and slowing down three times per revolution. This will in turn result in a cyclic so-called inertial torque being overlaid onto the combustion torque output from the crankshaft, which also has three pulses per revolution.

It is still a marked improvement on the inline four, whose pistons stop and start all at the same time, and is an improvement the inline-six shares with the inline-three - it is also the motivation for Yamaha's adoption of the cross-plane crank in its YZF-R1 motorcycle, with four unique piston phases. Inertial torque is typically only of concern in extreme situations, i.e. high reciprocating mass and high engine speed. It can however affect low-speed running and flywheel sizing in any case.

Two-stroke

An even-firing six cylinder two-stroke engine requires ignitions at 60° intervals or else it would run with simultaneous ignitions and be no smoother than a triple in power delivery. As such, it also requires crank throws at 60° - such designs appear to have been limited to Diesels such as the Detroit 71 series and some marine engines, as well as outboard motors.

Some of the 120 possible crankshaft configurations have useful properties, but all of them have a rocking imbalance of some kind which may or may not require a balance shaft, depending on the application. This is because the six pistons with six unique phases cannot be "paired" as in the four-stroke case. The Detroit engines used a configuration that, once the primary rocking couple was balanced out, was also perfectly balanced at all other rocking couples until 6th-order. Mercury came to use a configuration that canceled only the primary rocking couple and was run without a balancer.

The reciprocating masses of all configurations are still only imbalanced at 6th-order and up in their plane of motion, but the balance of kinetic energy exchange between pistons has improved to a residual 6th-order-and-up inertial torque oscillation compared with the four-stroke design being imbalanced at 3rd-order and up.

Crankshaft Design

Crankshaft with 4 main bearings.

Crankshafts on six-cylinder engines generally have either four or seven main bearings. Larger engines and diesels tend to use seven bearings because of high loadings and to avoid crankshaft flex. Because of the six-cylinder engine's smooth characteristic, there is a tendency for a driver to load the engine at low engine speeds. This can produce crankshaft flex in four main bearing designs where the crank spans the distance of two cylinders between main bearings. This distance is longer than the distance between two adjacent main bearings on a V6 with four mains, because the V6 has cylinder bores on opposite banks which overlap significantly; the overlap may be as high as 100%, minus the width of one connecting rod (1.00" or so). In addition, modern high-compression engines subject the crankshaft to greater bending loads from higher peak gas pressures, requiring the crankthrows to have greater support from adjacent bearings, so it is now customary to design straight-sixes with seven main bearings.

Many of the more sporty high-performance engines use the four bearing design because of better torsional stiffness (e.g., BMW small straight 6, Ford's Zephyr 6). The accumulated length of main bearing journals gives a relatively torsionally flexible crankshaft. The four main bearing design has only six crank throws and four main journals, so is much stiffer in the torsional domain. At high engine speeds, the lack of torsional stiffness can make the seven main bearing design susceptible to torsional flex and potential breakage. Another factor affecting large straight-six engines is the end-mounted timing chain which connects any camshafts to the crankshaft. The camshafts are also quite long and subject to torsional flex as they in turn operate valves alternately near the front of the engine and near the rear. At high engine speeds, camshafts can flex torsionally in addition to the crankshaft, contributing to valve timing for the cylinders furthest from the cam drive becoming inaccurate and erratic, losing power, and in extreme cases resulting in mechanical interference between valve and piston — with catastrophic results. Some designers have experimented with installing the timing chain/gears in the middle of the engine (between cylinders 3 and 4) or adding a second timing chain at the rear of the engine. Either method can solve the problem at the cost of additional complexity.

Another factor reducing the ability of the large six-cylinder engines to achieve high speed is the simple geometric reality of a relatively long stroke (undersquare) design. A straight-six is a long engine, and the designer is usually encouraged to make it as short as possible, while height is not usually a problem. Hence, the tendency to use a longer stroke and smaller bore than in a V engine to achieve a given capacity. By contrast, a long-stroke V engine tends to become too wide, which encourages increasing the bore rather than the stroke to increase displacement. The typically longer stroke of the straight-six increases crank throw and piston speed, and so tends to reduce the rpm rating of the engine.

V ENGINE

A V engine, or Vee engine is a common configuration for an internal combustion engine. The cylinders and pistons are aligned, in two separate planes or 'banks', so that they appear to be in a "V" when viewed along the axis of the crankshaft. The Vee configuration generally reduces the overall engine length, height and weight compared with an equivalent inline configuration.

The first V-type engine, a 2-cylinder vee twin, was built in 1889 by Daimler, to a design by Wilhelm Maybach. By 1903 V8 engines were being produced for motor boat racing by the Société Antoinette to designs by Léon Levavasseur, building on experience gained with in-line four-cylinder engines. In 1904, the Putney Motor Works completed a new V12, 150bhp 18.4 litre engine – the first V12 engine produced for any purpose. This one was manufactured for two Russian brothers making a dirigible. They ran out of money and Commander May bought it on a sale or return basis for Motor boat racing, having some moderate success in 1908. The engine was exposed and the hot coil ignition created misfiring on becoming wet with the spray. Robert Bosch supplied the very first magnetos and the problem was solved.

Characteristics

Usually, each pair of corresponding pistons from each bank of cylinders share one crankpin on the crankshaft, either by master/slave rods or by two ordinary connecting rods side by side. However, some V-twin engine designs have two-pin cranks, while other V configurations include split crankpins for more even firing.

V-engines are generally more compact than straight engines with cylinders of the same dimensions and number. This effect increases with the number of cylinders in the engine; there might be no noticeable difference in overall size between V-twin and straight-twin engines while V8 engines are much more compact than straight-eight engines.

Various cylinder bank angles of Vee are used in different engines; depending on the number of cylinders, there may be angles that work better than others for stability. Very narrow angles of Vee combine some of the advantages of the Vee engine and the straight engine (primarily in the form of compactness) as well as disadvantages; the concept is an old one pioneered by Lancia's V4 engine in the 1920s, but recently reworked by Volkswagen Group with their VR engines, which is actually a combination of V and straight configuration.

Some Vee configurations are well-balanced and smooth, while others are less smoothly running than their equivalent straight counterparts. V8s with crossplane crankshaft can be easily balanced with the use of counterweights only. V12s, being in effect two straight-6 engines married together, are fully balanced; if the V-angle is 60° for 4-stroke or 30° for 2-stroke, they also have even firing. Others, such as the V2, V4, V6, flatplane V8, V10, V14 and V18 engine show increased vibration and generally require balance shafts or split crankshafts.

Inverted Engines

Certain types of Vee engine have been built as inverted engines, most commonly for aircraft. Advantages include better visibility in a single-engined airplane, and lower centre of gravity. Examples include World War II German Daimler-Benz DB 601, Junkers Jumo, and Argus Motoren piston engines.

Specific Configurations

It is common practice for Vee engines to be described with **"V#"** notation, where # is how many cylinders it has:

- V2, commonly referred to as a V-twin
- V3
- V4
- V5
- V6
- V8
- V10
- V12

- V14

- V16

- V18

- V20

- V24

W ENGINE

A W engine is a type of reciprocating engine arranged with its cylinders in a configuration in which the cylinder banks resemble the letter W, in the same way those of a V engine resemble the letter V.

Four different configurations have been called W engines:

- Three banks of cylinders sharing a common crankshaft, a configuration also known as broad arrow configuration due to its shape resembling the British government broad arrow property mark.

- Four banks of cylinders sharing a common crankshaft, also called a 'double-V'.

- Two banks of cylinders with two crankshafts.

- Four banks of cylinders with two crankshafts.

The Original "Three-bank" Design

Napier Lion VII.

The classical W engine uses three banks of cylinders, all connected to one crankshaft.

One of the first W engines was the Anzani 3-cylinder, built in 1906, to be used in Anzani motorcycles. It is this W3 engine which also powered the Blériot XI, the aircraft used by Louis Blériot when, on 25 July 1909, he made the first flight across the English Channel. Shortly afterward the

W3 configuration was changed to a 120°-angle, three-cylinder radial engine configuration as the original W3 engine's replacement.

A three-bank W12 design was also pursued by Audi, which later abandoned the project. Volkswagen Group built an experimental W18 engine for Bugatti's EB 118 and EB 218 concept cars, but the design was determined to be impractical because of the irregular firing order required by the three rows of six cylinders.

The Feuling W

Similar to the W3 built by Anzani in 1906, the Feuling W3 is a 180 horsepower (134 kW; 182 PS), three-cylinder air-cooled engine for motorcycle cruisers. Like radial aircraft engines it has a master connecting rod and two slave rods connected to the three 101.6-millimetre (4.00 in) pistons. Motorcycle Cruiser Magazine reviewed Feulings's "Warlock" powered motorcycle in the October edition of 2000. Cory Ness built his chopper using a Feuling W3 engine during a Biker Build Off episode.

The Modern "four-bank" Design

A W16 engine from the Bugatti Veyron.

Volkswagen Group created the first successful automotive W engine, with the introduction of its W8 (as a testbed for the W12). The W12 combines two narrow-angle VR6 engine cylinder heads around a single crankshaft for a total of four banks of cylinders. For this reason, the four-bank configuration is sometimes, and more accurately, referred to as a "VV" ("vee-vee" or "double-vee") or "WR", to distinguish it from the traditional three-bank "W" design (the earlier W8 combined two VR4 engines).

The W8 was used in the B5.5 Volkswagen Passat and the W12 is used in the Volkswagen Phaeton, the Volkswagen Touareg, the Audi A8, and the Bentley Continental GT — though in the latter application, the engine has been highly modified by Bentley, and fitted with twin turbochargers. As a result, it produces considerably more power than the original version. The narrow (15°) angle between bank pairs makes this resemble a V12 engine, in that it has just two cylinder heads and two sets of camshafts. The W12 engine has bore-stroke of 84.0 millimetres (3.31 in) and 90.2 millimetres (3.55 in).

Volkswagen Group went on to produce a W16 engine prototype which produced 465 kilowatts (632 PS; 624 bhp) for the Bentley Hunaudières concept car. A quad-turbocharged version of this engine went into production in 2005 powering the 736 kilowatts (1,001 PS; 987 bhp) Bugatti Veyron

EB16.4. The major advantage of these engines is packaging; they contain high numbers of cylinders but are relatively compact in their external dimensions.

The W-engine in the Bugatti Veyron

In 2006, the Volkswagen Group-owned Bugatti produced the Bugatti Veyron EB16.4 with an 8.0 litre W16 engine. This has four turbochargers, and it produces DIN rated motive power output of 736 kilowatts (1,001 PS; 987 bhp) at 6,000 revolutions per minute (rpm). It uses four valves per cylinder, 64 valves total, with four overhead camshafts arranged in a double overhead camshaft (two overhead camshafts per cylinder bank) layout, and a bore-stroke ratio 1:1 (both bore and stroke are 86.0 millimetres or 3.39 inches).

DIESEL ENGINE

The diesel engine (also known as a compression-ignition or CI engine), named after Rudolf Diesel, is an internal combustion engine in which ignition of the fuel is caused by the elevated temperature of the air in the cylinder due to the mechanical compression (adiabatic compression). This contrasts with spark-ignition engines such as a petrol engine (gasoline engine) or gas engine (using a gaseous fuel as opposed to petrol), which use a spark plug to ignite an air-fuel mixture.

Diesel engines work by compressing only the air. This increases the air temperature inside the cylinder to such a high degree that atomised diesel fuel injected into the combustion chamber ignites spontaneously. With the fuel being injected into the air just before combustion, the dispersion of the fuel is uneven; this is called a heterogeneous air-fuel mixture. The torque a diesel engine produces is controlled by manipulating the air ratio; instead of throttling the intake air, the diesel engine relies on altering the amount of fuel that is injected, and the air ratio is usually high.

The diesel engine has the highest thermal efficiency (engine efficiency) of any practical internal or external combustion engine due to its very high expansion ratio and inherent lean burn which enables heat dissipation by the excess air. A small efficiency loss is also avoided compared to two-stroke non-direct-injection gasoline engines since unburned fuel is not present at valve overlap and therefore no fuel goes directly from the intake/injection to the exhaust. Low-speed diesel engines (as used in ships and other applications where overall engine weight is relatively unimportant) can reach effective efficiencies of up to 55%.

The world's largest diesel engines put in service are 14-cylinder, two-stroke watercraft diesel engines; they produce a peak power of almost 100 MW each.

Operating Principle

Characteristics

The characteristics of a diesel engine are:

- Compression ignition: Due to almost adiabatic compression, the fuel ignites without any ignition-initiating apparatus such as spark plugs.

- Mixture formation inside the combustion chamber: Air and fuel are mixed in the combustion chamber and not in the inlet manifold.

- Engine speed adjustment solely by mixture quality: Instead of throttling the air-fuel mixture, the amount of torque produced (resulting in crankshaft rotational speed differences) is set solely by the mass of injected fuel, always mixed with as much air as possible.

- Heterogeneous air-fuel mixture: The dispersion of air and fuel in the combustion chamber is uneven.

- High air ratio: Due to always running on as much air as possible and not depending on exact mixture of air and fuel, diesel engines have an air-fuel ratio leaner than stochiometric ($\lambda_v \geq \lambda_{min} > 1$).

- Diffusion flame: At combustion, oxygen first has to diffuse into the flame, rather than having oxygen and fuel already mixed before combustion, which would result in a premixed flame.

- Fuel with high ignition performance: As diesel engines solely rely on compression ignition, fuel with high ignition performance (cetane rating) is ideal for proper engine operation, fuel with a good knocking resistance (octane rating), e.g. petrol, is suboptimal for diesel engines.

Cycle of the Diesel Engine

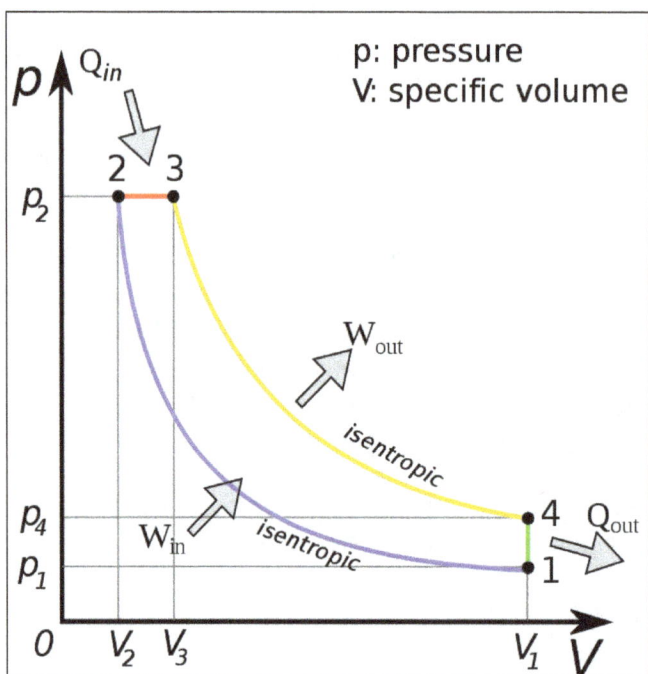

p-V Diagram for the ideal diesel cycle. The cycle follows the numbers 1–4 in clockwise direction. The horizontal axis is volume of the cylinder. In the diesel cycle the combustion occurs at almost constant pressure. On this diagram the work that is generated for each cycle corresponds to the area within the loop.

Diesel engine model, left side.

The diesel internal combustion engine differs from the gasoline powered Otto cycle by using highly compressed hot air to ignite the fuel rather than using a spark plug (compression ignition rather than spark ignition).

Diesel engine model, right side.

In the diesel engine, only air is initially introduced into the combustion chamber. The air is then compressed with a compression ratio typically between 15:1 and 23:1. This high compression causes the temperature of the air to rise. At about the top of the compression stroke, fuel is injected directly into the compressed air in the combustion chamber. This may be into a (typically toroidal) void in the top of the piston or a pre-chamber depending upon the design of the engine. The fuel injector ensures that the fuel is broken down into small droplets, and that the fuel is distributed evenly. The heat of the compressed air vaporises fuel from the surface of the droplets. The vapour is then ignited by the heat from the compressed air in the combustion chamber, the droplets continue to vaporise from their surfaces and burn, getting smaller, until all the fuel in the droplets has been burnt. Combustion occurs at a substantially constant pressure during the initial part of the power stroke. The start of vaporisation causes a delay before ignition and the characteristic diesel knocking sound as the vapour reaches ignition temperature and causes an abrupt increase in pressure above the piston. When combustion is complete the combustion gases expand as the piston descends further; the high pressure in the cylinder drives the piston downward, supplying power to the crankshaft.

As well as the high level of compression allowing combustion to take place without a separate ignition system, a high compression ratio greatly increases the engine's efficiency. Increasing the compression ratio in a spark-ignition engine where fuel and air are mixed before entry to the cylinder is limited by the need to prevent damaging pre-ignition. Since only air is compressed in a diesel engine, and fuel is not introduced into the cylinder until shortly before top dead centre (TDC), premature detonation is not a problem and compression ratios are much higher.

The p–V diagram is a simplified and idealised representation of the events involved in a diesel engine cycle, arranged to illustrate the similarity with a Carnot cycle. Starting at 1, the piston is at bottom dead centre and both valves are closed at the start of the compression stroke; the cylinder contains air at atmospheric pressure. Between 1 and 2 the air is compressed adiabatically – that is without heat transfer to or from the environment – by the rising piston. (This is only approximately true since there will be some heat exchange with the cylinder walls.) During this compression, the volume is reduced, the pressure and temperature both rise. At or slightly before 2 (TDC) fuel is injected and burns in the compressed hot air. Chemical energy is released and this constitutes an injection of thermal energy (heat) into the compressed gas. Combustion and heating occur between 2 and 3. In this interval the pressure remains constant since the piston descends, and the volume increases; the temperature rises as a consequence of the energy of combustion. At 3 fuel injection and combustion are complete, and the cylinder contains gas at a higher temperature than at 2. Between 3 and 4 this hot gas expands, again approximately adiabatically. Work is done on the system to which the engine is connected. During this expansion phase the volume of the gas rises, and its temperature and pressure both fall. At 4 the exhaust valve opens, and the pressure falls abruptly to atmospheric (approximately). This is unresisted expansion and no useful work is done by it. Ideally the adiabatic expansion should continue, extending the line 3–4 to the right until the pressure falls to that of the surrounding air, but the loss of efficiency caused by this unresisted expansion is justified by the practical difficulties involved in recovering it (the engine would have to be much larger). After the opening of the exhaust valve, the exhaust stroke follows, but this (and the following induction stroke) are not shown on the diagram. If shown, they would be represented by a low-pressure loop at the bottom of the diagram. At 1 it is assumed that the exhaust and induction strokes have been completed, and the cylinder is again filled with air. The piston-cylinder system absorbs energy between 1 and 2 – this is the work needed to compress the air in the cylinder, and is provided by mechanical kinetic energy stored in the flywheel of the engine. Work output is done by the piston-cylinder combination between 2 and 4. The difference between these two increments of work is the indicated work output per cycle, and is represented by the area enclosed by the p–V loop. The adiabatic expansion is in a higher pressure range than that of the compression because the gas in the cylinder is hotter during expansion than during compression. It is for this reason that the loop has a finite area, and the net output of work during a cycle is positive.

Efficiency

Due to its high compression ratio, the diesel engine has a high efficiency, and the lack of a throttle valve means that the charge-exchange losses are fairly low, resulting in a low specific fuel consumption, especially in medium and low load situations. This makes the diesel engine very economical. Even though diesel engines have a theoretical efficiency of 75%, in practice it is much lower. In his 1893 essay Theory and Construction of a Rational Heat Motor, Rudolf Diesel describes that the effective efficiency of the diesel engine would be in between 43.2% and 50.4%, or maybe even greater. Modern passenger car diesel engines may have an effective efficiency of up to 43%, whilst engines in large diesel trucks, and buses can achieve peak efficiencies around

45%. However, average efficiency over a driving cycle is lower than peak efficiency. For example, it might be 37% for an engine with a peak efficiency of 44%. The highest diesel engine efficiency of up to 55% is achieved by large two-stroke watercraft diesel engines.

Major Advantages

Diesel engines have several advantages over engines operating on other principles:

- The diesel engine has the highest effective efficiency of all combustion engines.

    ◦ Diesel engines inject the fuel directly into the combustion chamber, have no intake air restrictions apart from air filters and intake plumbing and have no intake manifold vacuum to add parasitic load and pumping losses resulting from the pistons being pulled downward against intake system vacuum. Cylinder filling with atmospheric air is aided and volumetric efficiency is increased for the same reason.

    ◦ Although the fuel efficiency (mass burned per energy produced) of a diesel engine drops at lower loads, it doesn't drop quite as fast as that of a typical petrol or turbine engine.

- Diesel engines can combust a huge variety of fuels, including several fuel oils, that have advantages over fuels such as petrol. These advantages include:

    ◦ Low fuel costs, as fuel oils are relatively cheap.

    ◦ Good lubrication properties.

    ◦ High energy density.

    ◦ Low risk of catching fire, as they do not form a flammable vapour.

    ◦ Biodiesel is an easily synthesised, non-petroleum-based fuel (through transesterification) which can run directly in many diesel engines, while gasoline engines either need adaptation to run synthetic fuels or else use them as an additive to gasoline (e.g., ethanol added to gasohol).

- Diesel engines have a very good exhaust-emission behaviour. The exhaust contains minimal amounts of carbon monoxide and hydrocarbons. Direct injected diesel engines emit approximately as much nitrogen oxide as Otto cycle engines. Swirl chamber and precombustion chamber injected engines, however, emit approximately 50% less nitrogen oxide than Otto cycle engines when running under full load. Compared with Otto cycle engines, diesel engines emit 10 times less pollutants and 3 times less carbon dioxide.

- They have no high voltage electrical ignition system, resulting in high reliability and easy adaptation to damp environments. The absence of coils, spark plug wires, etc., also eliminates a source of radio frequency emissions which can interfere with navigation and communication equipment, which is especially important in marine and aircraft applications, and for preventing interference with radio telescopes. (For this reason, only diesel-powered vehicles are allowed in parts of the American National Radio Quiet Zone).

- Diesel engines can accept super or turbocharging pressure without any natural limit, constrained only by the design and operating limits of engine components, such as pressure, speed and load. This is unlike petrol engines, which inevitably suffer detonation at higher pressure if engine tuning and fuel octane adjustments are not made to compensate.

Fuel Injection

Diesel engines rely on internal mixture formation, which means that they require a fuel injection system. The fuel is injected directly into the combustion chamber, which can be either a segmented combustion chamber or an unsegmented combustion chamber. Fuel injection with the latter is referred to as direct injection (DI), whilst injection into the former is called indirect injection (IDI). In diesel engine terminology, indirect injection does not mean fuel injection into the inlet manifold or anywhere else outside the cylinder or combustion chamber: in fact, the definition of the diesel engine excludes such injection methods. For creating the fuel pressure, diesel engines usually have an injection pump. There are several different types of injection pumps and methods for creating a fine air-fuel mixture. Over the years many different injection methods have been used. These can be described as the following:

- Air blast, where the fuel is blown into the cylinder by a blast of air.

- Solid fuel/hydraulic injection, where the fuel is pushed through a spring loaded valve/injector to produce a combustible mist.

- Mechanical unit injector, where the injector is directly operated by a cam and fuel quantity is controlled by a rack or lever.

- Mechanical electronic unit injector, where the injector is operated by a cam and fuel quantity is controlled electronically.

- Common rail mechanical injection, where fuel is at high pressure in a common rail and controlled by mechanical means.

- Common rail electronic injection, where fuel is at high pressure in a common rail and controlled electronically.

Torque Controlling

Due to the way diesel engines work, a vital component of all diesel engines is a mechanical or electronic governor which regulates the torque of the engine and thus idling speed and maximum speed by controlling the rate of fuel delivery. This means a change of λ_v. Unlike Otto-cycle engines, incoming air is not throttled. Mechanically governed fuel injection systems are driven by the engine's gear train. These systems use a combination of springs and weights to control fuel delivery relative to both load and speed. Modern electronically controlled diesel engines control fuel delivery by use of an electronic control module (ECM) or electronic control unit (ECU). The ECM/ECU receives an engine speed signal, as well as other operating parameters such as intake manifold pressure and fuel temperature, from a sensor and controls the amount of fuel and start of injection timing through actuators to maximise power and efficiency and minimise emissions. Controlling the timing of the start of injection of fuel into the cylinder is a key to minimizing emissions, and maximizing fuel economy (efficiency), of the engine. The timing is measured in degrees of crank angle of the piston before top dead centre. For example, if the ECM/ECU initiates fuel injection when the piston is 10° before TDC, the start of injection, or timing, is said to be 10° before TDC. Optimal timing will depend on the engine design as well as its speed and load.

Types of Fuel Injection

Air-blast Injection

Typical early 20th century air-blast injected diesel engine, rated at 59 kW.

Diesel's original engine injected fuel with the assistance of compressed air, which atomised the fuel and forced it into the engine through a nozzle (a similar principle to an aerosol spray). The nozzle opening was closed by a pin valve lifted by the camshaft to initiate the fuel injection before top dead centre (TDC). This is called an air-blast injection. Driving the compressor used some power but the efficiency was better than the efficiency of any other combustion engine at that time. Also, air-blast injection made engines very clunky and heavy and did not allow for quick load alteration, thus rendering it unusable for road vehicles.

Indirect Injection

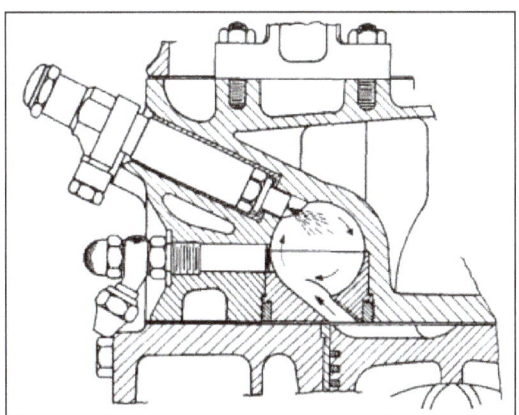

Ricardo Comet indirect injection chamber.

An indirect diesel injection system (IDI) engine delivers fuel into a small chamber called a swirl chamber, precombustion chamber, pre chamber or ante-chamber, which is connected to the cylinder by a narrow air passage. Generally the goal of the pre chamber is to create increased turbulence for better air/fuel mixing. This system also allows for a smoother, quieter running engine, and because fuel mixing is assisted by turbulence, injector pressures can be lower. Most IDI systems use a single orifice injector. The pre-chamber has the disadvantage of lowering efficiency due to increased heat loss to the engine's cooling system, restricting the combustion burn, thus reducing the efficiency by 5–10%. IDI engines are also more difficult to start and usually require the use of glow plugs. IDI engines may be cheaper to build but generally require a higher compression ratio than the DI counterpart. IDI also makes it easier to produce smooth, quieter running engines with a simple mechanical injection system since exact injection timing is not as critical. Most modern automotive engines are DI which have the benefits of greater efficiency and easier starting; however, IDI engines can still be found in the many ATV and small diesel applications. Indirect injected diesel engines use pintle-type fuel injectiors.

Helix-controlled Direct Injection

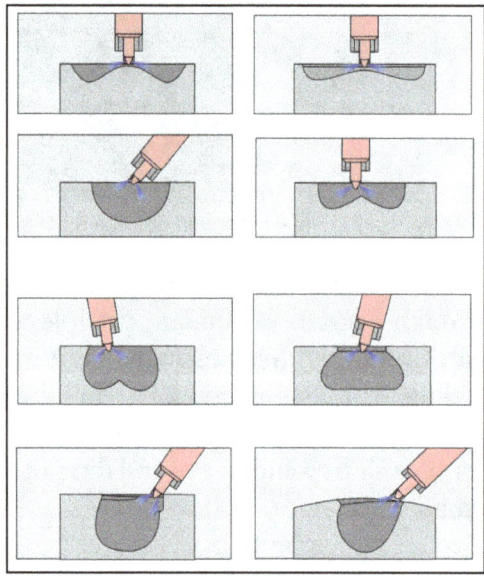

Different types of piston bowls.

Direct injection Diesel engines inject fuel directly into the cylinder. Usually there is a combustion cup in the top of the piston where the fuel is sprayed. Many different methods of injection can be used. Usually, an engine with helix-controlled mechanic direct injection has either an inline or a distributor injection pump. For each engine cylinder, the corresponding plunger in the fuel pump measures out the correct amount of fuel and determines the timing of each injection. These engines use injectors that are very precise spring-loaded valves that open and close at a specific fuel pressure. Separate high-pressure fuel lines connect the fuel pump with each cylinder. Fuel volume for each single combustion is controlled by a slanted groove in the plunger which rotates only a few degrees releasing the pressure and is controlled by a mechanical governor, consisting of weights rotating at engine speed constrained by springs and a lever. The injectors are held open by the fuel pressure. On high-speed engines the plunger pumps are together in one unit. The length of fuel lines from the pump to each injector is normally the same for each cylinder in order to obtain the same pressure delay. Direct injected diesel engines usually use orifice-type fuel injectors.

Electronic control of the fuel injection transformed the direct injection engine by allowing much greater control over the combustion.

Unit Direct Injection

Unit direct injection, also known as Pumpe-Düse (pump-nozzle), is a high pressure fuel injection system that injects fuel directly into the cylinder of the engine. In this system the injector and the pump are combined into one unit positioned over each cylinder controlled by the camshaft. Each cylinder has its own unit eliminating the high-pressure fuel lines, achieving a more consistent injection. Under full load, the injection pressure can reach up to 220 MPa. Unit injection systems used to dominate the commercial diesel engine market, but due to higher requirements of the flexibility of the injection system, they have been rendered obsolete by the more advanced common-rail-system.

Common Rail Direct Injection

Common rail (CR) direct injection systems, unlike other injection systems, do not have a combined pressure creation and injection apparatus. A high-pressure injection pump creates a constant pressure, not depending on the engine speed or fuel mass injected. A buffer, the so-called rail, saves this pressure. This allows fuel injection at any given moment, even multiple injections in a very short amount of time. The Electronic Diesel Control unit (EDC) controls both rail pressure and injections depending on several different parameters of the engine. The injectors of older CR systems have solenoid-driven plungers for lifting the injection needle, whilst newer CR injectors use plungers driven by piezoelectric actuators, that have fewer moving mass and therefore allow even more injections in a very short period of time. The injection pressure of modern CR systems ranges from 140 MPa to 270 MPa.

Types

There are several different ways of categorising diesel engines, based on different design characteristics:

By Power Output

- Small <188 kW (252 hp).

- Medium 188–750 kW.

- Large >750 kW.

By Cylinder Bore

- Passenger car engines: 75...100 mm.

- Lorry and commercial vehicle engines: 90...170 mm.

- High-performance high-speed engines: 165...280 mm.

- Medium-speed engines: 240...620 mm.

- Low-speed two-stroke engines: 260...900 mm.

By Number of Strokes

- Four-stroke cycle.

- Two-stroke cycle.

By Piston and Connecting Rod

- Crosshead piston.

- Double-acting piston.

- Opposed piston.

- Trunk piston.

By Cylinder Arrangement

Regular cylinder configurations such as straight (inline), V, and boxer (flat) configurations can be used for diesel engines. The inline-six-cylinder design is the most prolific in light- to medium-duty engines, though inline-four engines are also common. Small-capacity engines (generally considered to be those below five litres in capacity) are generally four- or six-cylinder types, with the four-cylinder being the most common type found in automotive uses. The V configuration used to be common for commercial vehicles, but it has been abandoned in favour of the inline configuration.

By Engine Speeds

Günter Mau categorises diesel engines by their rotational speeds into three groups:

- High-speed engines (> 1,000 rpm).

- Medium-speed engines (300–1,000 rpm).

- Slow-speed engines (< 300 rpm).

High-speed Engines

High-speed engines are used to power trucks (lorries), buses, tractors, cars, yachts, compressors, pumps and small electrical generators. As of 2018, most high-speed engines have direct injection. Many modern engines, particularly in on-highway applications, have common rail direct injection. On bigger ships, high-speed diesel engines are often used for powering electric generators. The highest power output of high-speed diesel engines is approximately 5 MW.

Medium-speed Engines

Medium-speed engines are used in large electrical generators, ship propulsion and mechanical drive applications such as large compressors or pumps. Medium speed diesel engines operate on either diesel fuel or heavy fuel oil by direct injection in the same manner as low-speed engines. Usually, they are four-stroke engines with trunk pistons.

The power output of medium-speed diesel engines can be as high as 21,870 kW, with the effective efficiency being around 47-48% (1982). Most larger medium-speed engines are started with compressed air direct on pistons, using an air distributor, as opposed to a pneumatic starting motor acting on the flywheel, which tends to be used for smaller engines.

Medium-speed engines intended for marine applications are usually used to power (ro-ro) ferries, passenger ships or small freight ships. Using medium-speed engines reduces the cost of smaller ships and increases their transport capacity. In addition to that, a single ship can use two smaller engines instead of one big engine, which increases the ship's safety.

Low-speed Engines

The MAN B&W 5S50MC 5-cylinder, 2-stroke, low-speed marine diesel engine.
This particular engine is found aboard a 29,000 tonne chemical carrier.

Low-speed diesel engines are usually very large in size and mostly used to power ships. There are two different types of low-speed engines that are commonly used: Two-stroke engines with a crosshead, and four-stroke engines with a regular trunk-piston. Two-stroke engines have a limited rotational frequency and their charge exchange is more difficult, which means that they are usually bigger than four-stroke engines and used to directly power a ship's propeller. Four-stroke engines on ships are usually used to power an electric generator. An electric motor powers the propeller. Both types are usually very undersquare. Low-speed diesel engines (as used in ships and other applications where overall engine weight is relatively unimportant) often have an effective efficiency of up to 55%. Like medium-speed engines, low-speed engines are started with compressed air, and they use heavy oil as their primary fuel.

Two-stroke Engines

Two-stroke diesel engines use only two strokes instead of four strokes for a complete engine cycle. Filling the cylinder with air and compressing it takes place in one stroke, and the power and exhaust strokes are combined. The compression in a two-stroke diesel engine is similar to the compression that takes place in a four-stroke diesel engine: As the piston passes through bottom centre and starts upward, compression commences, culminating in fuel injection and ignition. Instead of a full set of valves, two-stroke diesel engines have simple intake ports, and exhaust ports (or exhaust valves). When the piston approaches bottom dead centre, both the intake and the exhaust ports are "open", which means that there is atmospheric pressure inside the cylinder. Therefore, some sort of pump is required to blow the air into the cylinder and the combustion gasses into the exhaust. This process is called scavenging. The pressure required is approximately 10-30 kPa.

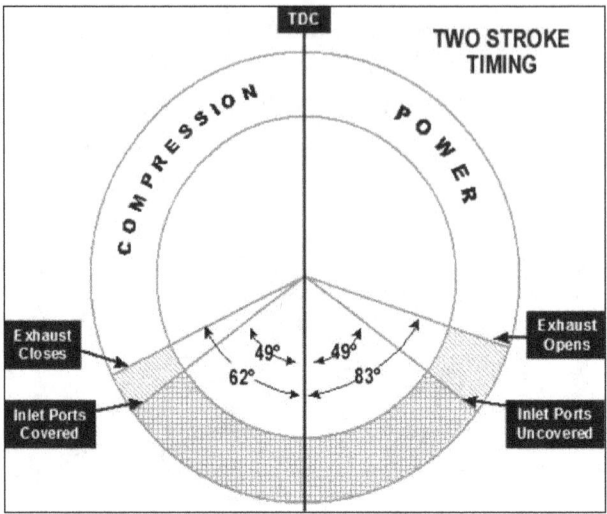

Detroit Diesel timing.

Scavenging

In general, there are three types of scavenging possible:

- Uniflow scavenging.

- Crossflow scavenging.

- Reverse flow scavenging.

Crossflow scavenging is incomplete and limits the stroke, yet some manufacturers used it. Reverse flow scavenging is a very simple way of scavenging, and it was popular amongst manufacturers until the early 1980s. Uniflow scavenging is more complicated to make but allows the highest fuel efficiency; since the early 1980s, manufacturers such as MAN and Sulzer have switched to this system. It is standard for modern marine two-stroke diesel engines.

Dual-fuel Diesel Engines

So-called dual-fuel diesel engines or gas diesel engines burn two different types of fuel simultaneously, for instance, a gaseous fuel and diesel engine fuel. The diesel engine fuel auto-ignites due to compression ignition, and then ignites the gaseous fuel. Such engines do not require any type of spark ignition and operate similar to regular diesel engines.

Diesel Engine Particularities

Torque and Power

Torque is a force applied to a lever at a right angle multiplied by the lever length. This means that the torque an engine produces depends on the displacement of the engine and the force that the gas pressure inside the cylinder applies to the piston, commonly referred to as effective piston pressure:

$$M = p_e \cdot V_h \cdot \pi^{-1} \cdot i^{-1}$$

M - Torque [N·m]; p_e - Effective piston pressure [kN·m^{-2}]; V_h - Displacement [dm³]; i - Strokes.

Example:

- Engine A: effective piston pressure=570 kN·m^{-2}, displacement= 2.2 dm³, strokes= 4, torque= 100 N·m

$$570 \cdot 2.2 \cdot \pi^{-1} \cdot 4^{-1} \approx 100$$

Power is the quotient of work and time:

$$P = 2\pi n M$$

P - Power [W]; M - Torque [N·m]; n - Time (crankshaft speed) [s^{-1}]

which means:

$$P = 2\pi \cdot n_1 \cdot M \cdot 60^{-1}$$

P - Power [W]; M - Torque [N·m]; n_1 - Time (crankshaft speed) [min^{-1}]

Example:

- Engine A: Power≈ 44,000 W, torque= 100 N·m, time= 4200 min^{-1}

$$44,000 \approx 2 \cdot \pi \cdot 4200 \cdot 100 \cdot 60^{-1}$$

- Engine B: Power≈ 44,000 W, torque= 260 N·m, time= 1600 min^{-1}

$$44,000 \approx 2 \cdot \pi \cdot 1600 \cdot 260 \cdot 60^{-1}$$

This means, that increasing either torque or time will result in an increase in power. As the maximum rotational frequency of the diesel engine's crankshaft is usually in between 3500...5000 min^{-1} due to diesel principle limitations, the torque of the diesel engine must be great to achieve a high power, or, in other words, as the diesel engine cannot use a lot of time for achieving a certain amount of power, it has to perform more work (=produce more torque).

Mass

The average diesel engine has a poorer power-to-mass ratio than the Otto engine. This is because the diesel must operate at lower engine speeds. Due to the higher operating pressure inside the combustion chamber, which increases the forces on the parts due to inertial forces, the diesel engine needs heavier, stronger parts capable of resisting these forces, which results in an overall greater engine mass.

Emissions

As diesel engines burn a mixture of fuel and air, the exhaust therefore contains substances that consist of the same chemical elements, as fuel and air. The main elements of air are nitrogen (N_2) and oxygen (O_2), fuel consists of hydrogen (H_2) and carbon (C). Burning the fuel will result in the final stage of oxidation. An ideal diesel engine, (a hypothetical model that we use as an example), running on an ideal air-fuel mixture, produces an exhaust that consists of carbon dioxide (CO_2), water (H_2O), nitrogen (N_2), and the remaining oxygen (O_2). The combustion process in a real engine differs from an ideal engine's combustion process, and due to incomplete combustion, the exhaust contains additional substances, most notably, carbon monoxide (CO), diesel particulate matter (PM), and due to dissociation, nitrogen oxide (NO_x).

When diesel engines burn their fuel with high oxygen levels, this results in high combustion temperatures and higher efficiency, and particulate matter tends to burn, but the amount of NO_x pollution tends to increase. NO_x pollution can be reduced by recirculating a portion of an engine's exhaust gas back to the engine cylinders, which reduces the oxygen quantity, causing a reduction of combustion temperature, and resulting in fewer NO_x. To further reduce NO_x emissions, lean NO_x traps (LNTs) and SCR-catalysts can be used. Lean NO_x traps adsorb the nitrogen oxide and "trap" it. Once the LNT is full, it has to be "regenerated" using hydrocarbons. This is achieved by using a very rich air-fuel mixture, resulting in incomplete combustion. An SCR-catalyst converts nitrogen oxide using urea, which is injected into the exhaust stream, and catalytically converts the NO x into nitrogen (N_2) and water (H_2O). Compared with an Otto engine, the diesel engine produces approximately the same amount of NO_x, but some older diesel engines may have an exhaust that contains up to 50% less NO_x. However, Otto engines, unlike diesel engines, can use a three-way-catalyst, that converts most of the NO_x.

Noise

Typical diesel engine noise of a 1950s direct injected
two-cylinder diesel engine (MWM AKD 112 Z, in idle).

The distinctive noise of a diesel engine is variably called diesel clatter, diesel nailing, or diesel knock. Diesel clatter is caused largely by the way the fuel ignites; the sudden ignition of the diesel fuel when injected into the combustion chamber causes a pressure wave, resulting in an audible "knock". Engine designers can reduce diesel clatter through: indirect injection; pilot or pre-injection; injection timing; injection rate; compression ratio; turbo boost; and exhaust gas recirculation (EGR). Common rail diesel injection systems permit multiple injection events as an aid to noise reduction. Therefore, newer diesel engines do not knock anymore. Diesel fuels with a higher cetane rating are more likely to ignite and hence reduce diesel clatter.

Cold Weather Starting

In general, diesel engines do not require any starting aid. In cold weather however, some diesel engines can be difficult to start and may need preheating depending on the combustion chamber design. The minimum starting temperature that allows starting without pre-heating is 40 °C for precombustion chamber engines, 20 °C for swirl chamber engines, and 0 °C for direct injected engines. Smaller engines with a displacement of less than 1 litre per cylinder usually have glowplugs, whilst larger heavy-duty engines have flame-start systems.

In the past, a wider variety of cold-start methods were used. Some engines, such as Detroit Diesel engines used a system to introduce small amounts of ether into the inlet manifold to start combustion. Instead of glowplugs, some diesel engines are equipped with starting aid systems that change valve timing. The simplest way this can be done is with a decompression lever. Activating the decompression lever locks the outlet valves in a slight down position, resulting in the engine not having any compression and thus allowing for turning the crankshaft over without resistance. When the crankshaft reaches a higher speed, flipping the decompression lever back into its normal position will abruptly re-activate the outlet valves, resulting in compression – the flywheel's mass moment of inertia then starts the engine. Other diesel engines, such as the precombustion chamber engine XII Jv 170/240 made by Ganz & Co., have a valve timing changing system that is operated by adjusting the inlet valve camshaft, moving it into a slight "late" position. This will make the inlet valves open with a delay, forcing the inlet air to heat up when entering the combustion chamber.

Supercharging and Turbocharging

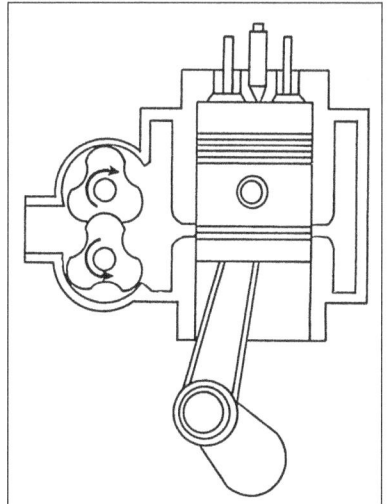

Two stroke diesel engine with Roots blower, typical of Detroit Diesel and some Electro-Motive Diesel Engines.

Turbocharged 1980s passenger car diesel engine with wastegate turbocharger and without intercooler (BMW M21).

As the diesel engine relies on manipulation of λ_v for torque controlling and speed regulation, the intake air mass does not have to precisely match the injected fuel mass (which would be $\lambda = 1$). Diesel engines are thus ideally suited for supercharging and turbocharging. An additional advantage of the diesel engine is the lack of fuel during the compression stroke. In diesel engines, the fuel is injected near top dead centre (TDC), when the piston is near its highest position. The fuel then ignites due to compression heat. Preignition, caused by the artificial turbocharger compression increase during the compression stroke, cannot occur.

Many diesels are therefore turbocharged and some are both turbocharged and supercharged. A turbocharged engine can produce more power than a naturally aspirated engine of the same configuration. A supercharger is powered mechanically by the engine's crankshaft, while a turbocharger is powered by the engine exhaust. Turbocharging can improve the fuel economy of diesel engines by recovering waste heat from the exhaust, increasing the excess air factor, and increasing the ratio of engine output to friction losses. Adding an intercooler to a turbocharged engine further increases engine performance by cooling down the air-mass and thus allowing more air-mass per volume.

A two-stroke engine does not have a discrete exhaust and intake stroke and thus is incapable of self-aspiration. Therefore, all two-stroke diesel engines must be fitted with a blower or some form of compressor to charge the cylinders with air and assist in dispersing exhaust gases, a process referred to as scavenging. Roots-type superchargers were used for ship engines until the mid-1950s, since 1955 they have been widely replaced by turbochargers. Usually, a two-stroke ship diesel engine has a single-stage turbocharger with a turbine that has an axial inflow and a radial outflow.

Fuel and Fluid Characteristics

In diesel engines, a mechanical injector system vaporises the fuel directly into the combustion chamber (as opposed to a Venturi jet in a carburetor, or a fuel injector in a manifold injection system vaporising fuel into the intake manifold or intake runners as in a petrol engine). This forced vaporisation means that less-volatile fuels can be used. More crucially, because only air is inducted into the cylinder in a diesel engine, the compression ratio can be much higher as there is no risk of pre-ignition provided the injection process is accurately timed. This means that cylinder temperatures are much higher in a diesel engine than a petrol engine, allowing less volatile fuels to be used.

The MAN 630's M-System diesel engine is a petrol engine (designed to run on NATO F 46/F 50 petrol), but it also runs on jet fuel, (NATO F 40/F 44), kerosene, (NATO F 58), and diesel engine fuel (NATO F 54/F 75).

Therefore, diesel engines can operate on a huge variety of different fuels. In general, fuel for diesel engines should have a proper viscosity, so that the injection pump can pump the fuel to the injection nozzles without causing damage to itself or corrosion of the fuel line. At injection, the fuel should form a good fuel spray, and it should not have a coking effect upon the injection nozzles. To ensure proper engine starting and smooth operation, the fuel should be willing to ignite and hence not cause a high ignition delay (this means that the fuel should have a high cetane number). Diesel fuel should also have a high lower heating value.

Inline mechanical injector pumps generally tolerate poor-quality or bio-fuels better than distributor-type pumps. Also, indirect injection engines generally run more satisfactorily on fuels with a high ignition delay (for instance, petrol) than direct injection engines. This is partly because an indirect injection engine has a much greater 'swirl' effect, improving vaporisation and combustion of fuel, and because (in the case of vegetable oil-type fuels) lipid depositions can condense on the cylinder walls of a direct-injection engine if combustion temperatures are too low (such as starting the engine from cold). Direct-injected engines with an MAN centre sphere combustion chamber rely on fuel condensing on the combustion chamber walls. The fuel starts vaporising only after ignition sets in, and it burns relatively smoothly. Therefore, such engines also tolerate fuels with poor ignition delay characteristics, and, in general, they can operate on petrol rated 86 RON.

Fuel Types

In his 1893 work Theory and Construction of a Rational Heat Motor, Rudolf Diesel considers using coal dust as fuel for the diesel engine. However, Diesel just considered using coal dust (as well as liquid fuels and gas); his actual engine was designed to operate on petroleum, which was soon replaced with regular petrol and kerosene for further testing purposes, as petroleum proved to be too viscous. In addition to kerosene and petrol, Diesel's engine could also operate on ligroin.

Before diesel engine fuel was standardised, fuels such as petrol, kerosene, gas oil, vegetable oil and mineral oil, as well as mixtures of these fuels, were used. Typical fuels specifically intended to be used for diesel engines were petroleum distillates and coal-tar distillates such as the following; these fuels have specific lower heating values of:

- Diesel oil: 10,200 kcal·kg⁻¹ (42.7 MJ·kg⁻¹) up to 10,250 kcal·kg⁻¹ (42.9 MJ·kg⁻¹).

- Heating oil: 10,000 kcal·kg⁻¹ (41.8 MJ·kg⁻¹) up to 10,200 kcal·kg⁻¹ (42.7 MJ·kg⁻¹).

- Coal-tar creosote: 9,150 kcal·kg⁻¹ (38.3 MJ·kg⁻¹) up to 9,250 kcal·kg⁻¹ (38.7 MJ·kg⁻¹).

- Kerosene: up to 10,400 kcal·kg⁻¹ (43.5 MJ·kg⁻¹).

The first diesel fuel standards were the DIN 51601, VTL 9140-001, and NATO F 54, which appeared after World War II. The modern European EN 590 diesel fuel standard was established in May 1993; the modern version of the NATO F 54 standard is mostly identical with it. The DIN 51628 biodiesel standard was rendered obsolete by the 2009 version of the EN 590; FAME biodiesel conforms to the EN 14214 standard. Watercraft diesel engines usually operate on diesel engine fuel that conforms to the ISO 8217 standard (Bunker C). Also, some diesel engines can operate on gasses (such as LNG).

Modern Diesel Fuel Properties

Modern diesel fuel properties		
	EN 590 (as of 2009)	EN 14214 (as of 2010)
Ignition performance	≥ 51 CN	≥ 51 CN
Density at 15 °C	820...845 kg·m⁻³	860...900 kg·m⁻³
Sulphur content	≤10 mg·kg⁻¹	≤10 mg·kg⁻¹
Water content	≤200 mg·kg⁻¹	≤500 mg·kg⁻¹
Lubricity	460 μm	460 μm
Viscosity at 40 °C	2.0...4.5 mm²·s⁻¹	3.5...5.0 mm²·s⁻¹
FAME content	≤7.0%	≥96.5%
Molar H/C ratio	–	1.69
Lower heating value	–	37.1 MJ·kg⁻¹

Gelling

DIN 51601 diesel fuel was prone to waxing or gelling in cold weather; both are terms for the solidification of diesel oil into a partially crystalline state. The crystals build up in the fuel system (especially in fuel filters), eventually starving the engine of fuel and causing it to stop running. Low-output electric heaters in fuel tanks and around fuel lines were used to solve this problem. Also, most engines have a spill return system, by which any excess fuel from the injector pump and injectors is returned to the fuel tank. Once the engine has warmed, returning warm fuel prevents waxing in the tank. Some manufacturers, such as BMW, recommended fuelling diesel cars with petrol to prevent the fuel from gelling when the temperatures dropped below −15 °C.

Safety

Fuel Flammability

Diesel fuel is less flammable than petrol, because its flash point is 55 °C, leading to a lower risk of fire caused by fuel in a vehicle equipped with a diesel engine.

Diesel fuel can create an explosive air/vapour mix under the right conditions. However, compared with petrol, it is less prone due to its lower vapour pressure, which is an indication of evaporation rate. The Material Safety Data Sheet for ultra-low sulfur diesel fuel indicates a vapour explosion hazard for diesel fuel indoors, outdoors, or in sewers.

Cancer

Diesel exhaust has been classified as an IARC Group 1 carcinogen. It causes lung cancer and is associated with an increased risk for bladder cancer.

Applications

The characteristics of diesel have different advantages for different applications.

Passenger Cars

Diesel engines have long been popular in bigger cars and have been used in smaller cars such as superminis in Europe since the 1980s. They were popular in larger cars earlier, as the weight and cost penalties were less noticeable. Smooth operation as well as high low end torque are deemed important for passenger cars and small commercial vehicles. The introduction of electronically controlled fuel injection significantly improved the smooth torque generation, and starting in the early 1990s, car manufacturers began offering their high-end luxury vehicles with diesel engines. Passenger car diesel engines usually have between three and ten cylinders, and a displacement ranging from 0.8 to 5.0 litres. Modern powerplants are usually turbocharged and have direct injection.

Diesel engines do not suffer from intake-air throttling, resulting in very low fuel consumption especially at low partial load (for instance: driving at city speeds). One fifth of all passenger cars worldwide have diesel engines, with many of them being in Europe, where approximately 47% of all passenger cars are diesel-powered. Daimler-Benz in conjunction with Robert Bosch GmbH produced diesel-powered passenger cars starting in 1936. The popularity of diesel-powered passenger cars in markets such as India, South Korea and Japan is increasing .

Commercial Vehicles and Lorries

In 1893, Rudolf Diesel suggested that the diesel engine could possibly power 'wagons' (lorries). The first lorries with diesel engines were brought to market in 1924.

Modern diesel engines for lorries have to be both extremely reliable and very fuel efficient. Common-rail direct injection, turbocharging and four valves per cylinder are standard. Displacements range from 4.5 to 15.5 litres, with power-to-mass ratios of 2.5–3.5 kg·kW^{-1} for heavy duty and 2.0–3.0 kg·kW^{-1} for medium duty engines. V6 and V8 engines used to be common, due to the relatively low engine mass the V configuration provides. Recently, the V configuration has been abandoned in favour of straight engines. These engines are usually straight-6 for heavy and medium duties and straight-4 for medium duty. Their under-square design causes lower overall piston speeds which results in increased lifespan of up to 1,200,000 km. Compared with 1970s diesel engines, the expected lifespan of modern lorry diesel engines has more than doubled.

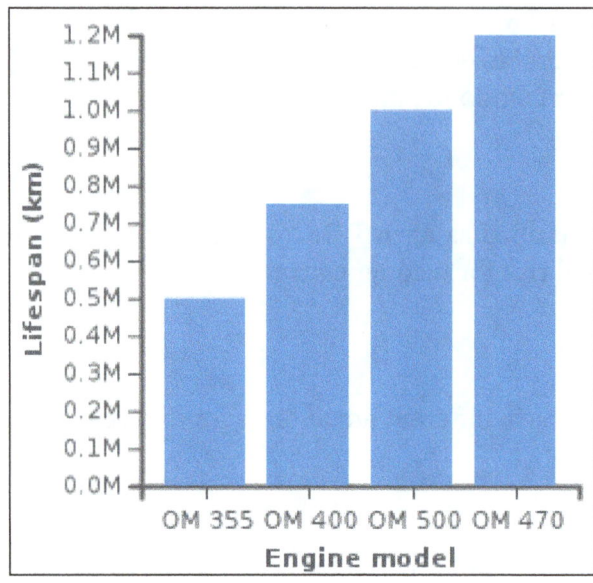

Lifespan of Mercedes-Benz diesel engines.

Railroad Rolling Stock

Diesel engines for locomotives are built for continuous operation and may require the ability to use poor quality fuel in some circumstances. Some locomotives use two-stroke diesel engines. Diesel engines have eclipsed steam engines as the prime mover on all non-electrified railroads in the industrialised world. The first diesel locomotives appeared in 1913, and diesel multiple units soon after. Many modern diesel locomotives are actually diesel-electric locomotives: the diesel engine is used to power an electric generator that in turn powers electric traction motors with no mechanical connection between diesel engine and traction. While electric locomotives have replaced the diesel locomotive for some passenger traffic in Europe and Asia, diesel is still today very popular for cargo-hauling freight trains and on tracks where electrification is not feasible.

In the 1940s, road vehicle diesel engines with power outputs of 150-200 PS (110-147 kW) were considered reasonable for DMUs. Commonly, regular truck powerplants were used. The height of these engines had to be less than 1,000 mm to allow underfloor installation. Usually, the engine was mated with a pneumatically operated mechanical gearbox, due to the low size, mass, and production costs of this design. Some DMUs used hydraulic torque converters instead. Diesel-electric transmission was not suitable for such small engines. In the 1930s, the Deutsche Reichsbahn standardised its first DMU engine. It was a 30.3 litre, 12-cylinder boxer unit, producing 275 PS (202 kW). Several German manufacturers produced engines according to this standard.

Watercraft

The requirements for marine diesel engines vary, depending on the application. For military use and medium-size boats, medium-speed four-stroke diesel engines are most suitable. These engines usually have up to 24 cylinders and come with power outputs in the one-digit Megawatt region. Small boats may use lorry diesel engines. Large ships use extremely efficient, low-speed two-stroke diesel engines. They can reach efficiencies of up to 55%. Unlike most regular diesel engines, two-stroke watercraft engines use highly viscous fuel oil. Submarines are usually diesel-electric.

One of the eight-cylinder 3200 I.H.P. Harland and Wolff – Burmeister & Wain diesel engines installed in the motorship Glenapp. This was the highest powered diesel engine yet (1920) installed in a ship. Note man standing lower right for size comparison.

The first diesel engines for ships were made by A. B. Diesels Motorer Stockholm in 1903. These engines were three-cylinder units of 120 PS (88 kW) and four-cylinder units of 180 PS (132 kW) and used for Russian ships. In World War I, especially submarine diesel engine development advanced quickly. By the end of the War, double acting piston two-stroke engines with up to 12,200 PS (9 MW) had been made for marine use.

Non-road Diesel Engines

Air-cooled diesel engine of a 1959 Porsche 218.

Non-road diesel engines are commonly used for construction equipment. Fuel efficiency, reliability and ease of maintenance are very important for such engines, whilst high power output and quiet operation are negligible. Therefore, mechanically controlled fuel injection and air-cooling are still very common. The common power outputs of non-road diesel engines vary a lot, with the smallest units starting at 3 kW, and the most powerful engines being heavy duty lorry engines.

Stationary Diesel Engines

Three English Electric 7SRL diesel-alternator sets being installed at the Saateni Power Station, Zanzibar 1955.

Stationary diesel engines are commonly used for electricity generation, but also for powering refrigerator compressors, or other types of compressors or pumps. Usually, these engines run permanently, either with mostly partial load, or intermittently, with full load. Stationary diesel engines powering electric generators that put out an alternating current, usually operate with alternating load, but fixed rotational frequency. This is due to the mains' fixed frequency of either 50 Hz (Europe), or 60 Hz (United States). The engine's crankshaft rotational frequency is chosen so that the mains' frequency is a multiple of it. For practical reasons, this results in crankshaft rotational frequencies of either 25 Hz (1500 per minute) or 30 Hz (1800 per minute).

Low Heat Rejection Engines

A special class of prototype internal combustion piston engines has been developed over several decades with the goal of improving efficiency by reducing heat loss. These engines are variously called adiabatic engines; due to better approximation of adiabatic expansion; low heat rejection engines, or high temperature engines. They are generally piston engines with combustion chamber parts lined with ceramic thermal barrier coatings. Some make use of pistons and other parts made of titanium which has a low thermal conductivity and density. Some designs are able to eliminate the use of a cooling system and associated parasitic losses altogether. Developing lubricants able to withstand the higher temperatures involved has been a major barrier to commercialization.

GASOLINE ENGINE

Gasoline Engine is an engine of class of internal-combustion engines that generate power by burning a volatile liquid fuel (gasoline or a gasoline mixture such as ethanol) with ignition initiated by an electric spark. Gasoline engines can be built to meet the requirements of practically any conceivable power-plant application, the most important being passenger automobiles, small trucks and buses, general aviation aircraft, outboard and small inboard marine units, moderate-sized stationary pumping, lighting plants, machine tools, and power tools. Four-stroke gasoline engines power the vast majority of automobiles, light trucks, medium-to-large motorcycles, and lawn mowers. Two-stroke gasoline engines are less common, but they are used for small outboard marine engines and in many handheld landscaping tools such as chain saws, hedge trimmers, and leaf blowers.

Types of Gasoline Engine

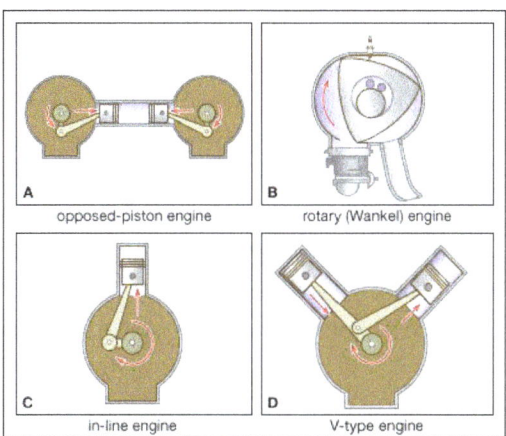

Four types of gasoline engines.

Gasoline engines can be grouped into a number of types depending on several criteria, including their application, method of fuel management, ignition, piston-and-cylinder or rotor arrangement, strokes per cycle, cooling system, and valve type and location. Here, they are described within the context of two basic engine types: piston-and-cylinder engines and rotary engines. In a piston-and-cylinder engine the pressure produced by combustion of gasoline creates a force on the head of a piston that moves the length of the cylinder in a reciprocating, or back-and-forth, motion. This force drives the piston away from the head of the cylinder and performs work. The rotary engine, also called the Wankel engine, does not have conventional cylinders fitted with reciprocating pistons. Instead, the gas pressure acts on the surfaces of a rotor, causing the rotor to turn and thus perform work.

Piston-and-cylinder Engines

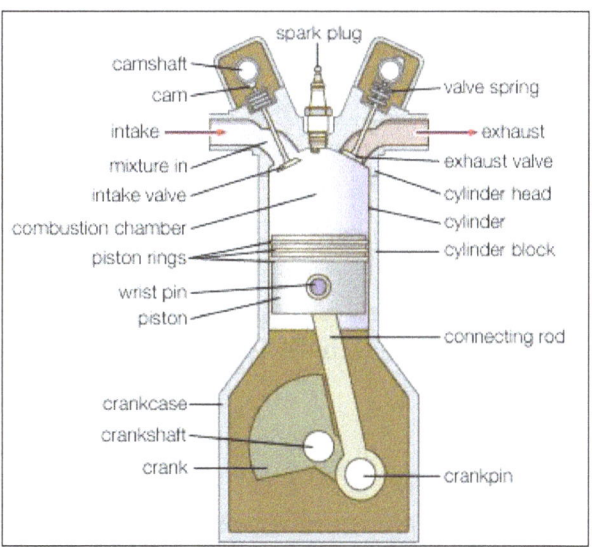

Typical piston-cylinder arrangement of a gasoline engine..

Most gasoline engines are of the reciprocating piston-and-cylinder type. The essential components of the piston-and-cylinder engine are shown in the figure. Almost all engines of this type follow either the four-stroke cycle or the two-stroke cycle.

Four-stroke Cycle

Of the different techniques for recovering the power from the combustion process, the most important so far has been the four-stroke cycle, a conception first developed in the late 19th century. The four-stroke cycle is illustrated in the figure. With the inlet valve open, the piston first descends on the intake stroke. An ignitable mixture of gasoline vapour and air is drawn into the cylinder by the partial vacuum thus created. The mixture is compressed as the piston ascends on the compression stroke with both valves closed. As the end of the stroke is approached, the charge is ignited by an electric spark. The power stroke follows, with both valves still closed and the gas pressure, due to the expansion of the burned gas, pressing on the piston head or crown. During the exhaust stroke the ascending piston forces the spent products of combustion through the open exhaust valve. The cycle then repeats itself. Each cycle thus requires four strokes of the piston—intake, compression, power, and exhaust—and two revolutions of the crankshaft.

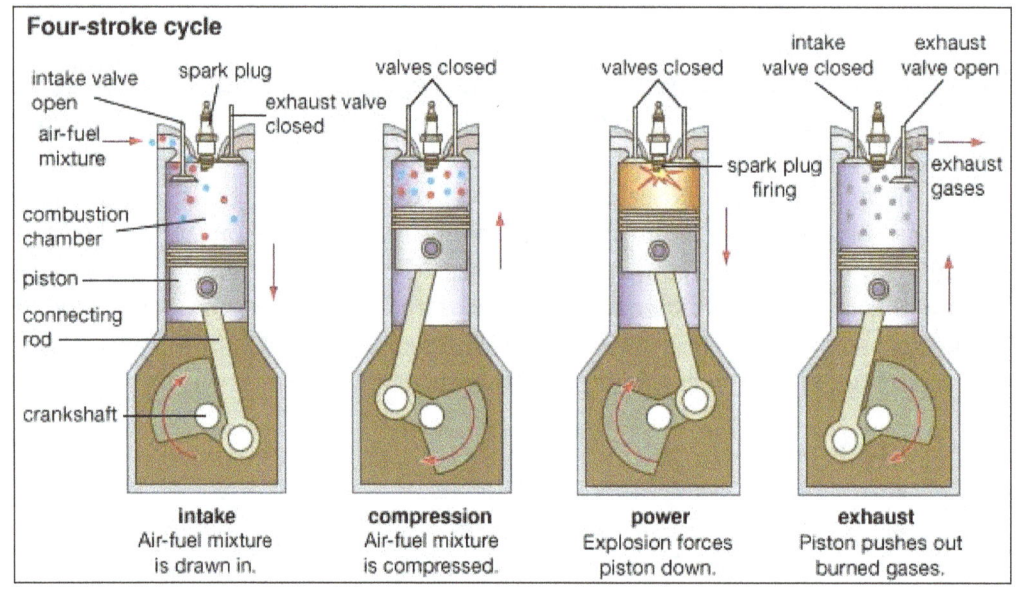

internal-combustion Engine: Four-stroke Cycle An internal-combustion engine goes through four strokes: intake, compression, combustion (power), and exhaust. As the piston moves during each stroke, it turns the crankshaft.

A disadvantage of the four-stroke cycle is that only half as many power strokes are completed as in the two-stroke cycle and only half as much power can be expected from an engine of a given size at a given operating speed. The four-stroke cycle, however, provides more positive clearing out of exhaust gases (scavenging) and reloading of the cylinders, reducing the loss of fresh charge to the exhaust.

Two-stroke Cycle

In the original two-stroke cycle (as developed in 1878), the compression and power stroke of the four-stroke cycle are carried out without the inlet and exhaust strokes, thus requiring only one revolution of the crankshaft to complete the cycle. The fresh fuel mixture is forced into the cylinder through circumferential ports by a rotary blower in the two-stroke-cycle engine of a so-called uniflow type. The exhaust gases pass through poppet valves in the cylinder head that are opened and closed by a cam-follower mechanism. The valves are timed to begin opening toward the end of the power stroke, after the cylinder pressure has dropped appreciably. The inlet ports in the cylinder wall start to uncover after the exhaust opening has decreased the cylinder pressure to the inlet pressure produced by the blower. The exhaust valves are allowed to remain open for a few degrees of crank rotation after the inlet ports have been covered by the rising piston on the compression stroke, thus allowing the persistency of flow to scavenge the cylinder more thoroughly. The compression and power strokes are similar to those of the four-stroke engine.

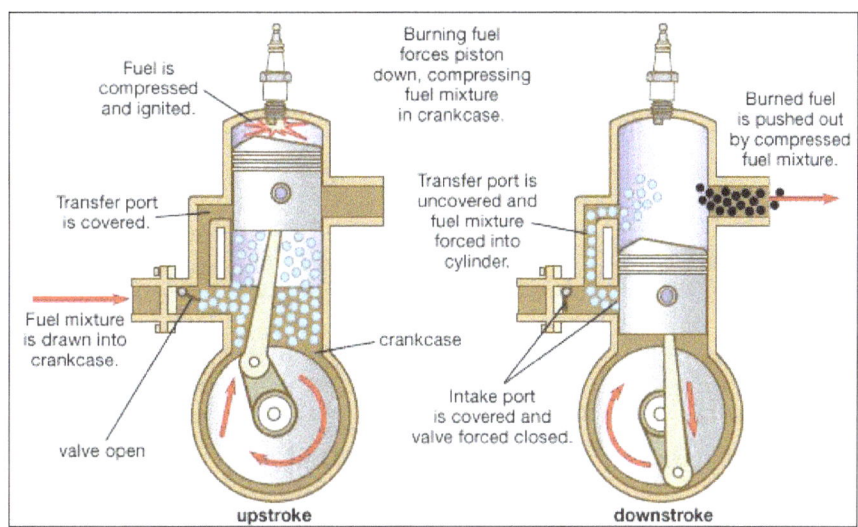

Blower-scavenged, two-stroke-cycle engine with uniflow scavenging.

A simplified version of the two-stroke-cycle engine was developed some years later (introduced in 1891) by using crankcase compression to pump the fresh charge into the cylinder. Instead of intake ports extending entirely around the lower cylinder wall, this engine has intake ports only halfway around; a second set of ports starts a little higher in the cylinder wall in the other half of the cylinder bore. These larger ports lead to the exhaust system. The inlet ports connect to a transfer passage leading to the fully enclosed crankcase. A spring-loaded inlet valve admits air into the crankcase on the upward, or compression, stroke of the piston. Air trapped in the crankcase is compressed by the descent of the piston on its power stroke. The piston thus uncovers the exhaust ports near the end of the power stroke, and slightly later it uncovers the inlet, or transfer, port on the opposite side of the cylinder to admit the compressed fresh mixture from the crankcase. The top face of the piston is designed to provide a deflector or baffle that directs the fresh load upward on the inlet side of the cylinder and then downward on the exhaust side, thus pushing the spent gases of the previous cycle out through the exhaust port on that side. This outflow continues after the inlet ports are covered by the rising piston on the compression stroke, until the exhaust ports are covered and compression of the fresh load begins. This loading process, called loop scavenging, is the simplest known method of replacing the exhaust products with a fresh mixture and creating a cycle with only compression and power strokes.

Such a system is used in many small gasoline engines (e.g., small outboard motors) and for gasoline-powered appliances. A disadvantage is that the return flow of the gases causes a slight loss of fresh charge through the exhaust ports. Because of this loss, carburetor engines operating on the two-stroke cycle lack the fuel economy of four-stroke engines. The loss can be avoided by equipping them with fuel-injection systems instead of carburetors and injecting the fuel directly into the cylinders after scavenging. Such an arrangement is attractive as a means of attaining high power output from a relatively small engine, and development of the turbocharger for this application holds promise of further improvement.

Opposed-piston Engines

The opposed-piston engine also provides uniflow scavenging. This engine has two pistons moving in opposite directions in the same cylinder. Two sets of ports extending entirely around the cylinder bore are located so that one set is covered and uncovered by one piston and the other set is controlled by the second piston. A second crankshaft, to which the upper pistons are attached, is located at the top of the engine, and the two shafts are connected by gears.

The opposed-piston design has two major advantages: Reciprocating masses move in opposite directions, providing excellent balance; and the poppet valves necessary in other uniflow-scavenged two-stroke-cycle engines are eliminated.

Rotary (Wankel) Engines

The rotary-piston internal-combustion engine developed in Germany is radically different in structure from conventional reciprocating piston engines. This engine was conceived by Felix Wankel, a specialist in the design of sealing devices, and experimental units were built and tested by a German firm beginning in 1956. Instead of pistons that move up and down in cylinders, the Wankel engine has an equilateral triangular orbiting rotor. The rotor turns in a closed chamber, and the three apexes of the rotor maintain a continuous sliding contact with the curved inner surface of the casing. The curve-sided rotor forms three crescent-shaped chambers between its sides and the curved wall of the casing. The volumes of the chambers vary with rotor position. Maximum volume is attained in each chamber when the side of the rotor forming it is parallel with the minor diameter of the casing; the volume is reduced to a minimum when the rotor side is parallel with the major diameter. Shallow pockets recessed in the flank of the rotor control the shape of the combustion chambers and establish the compression ratio of the engine.

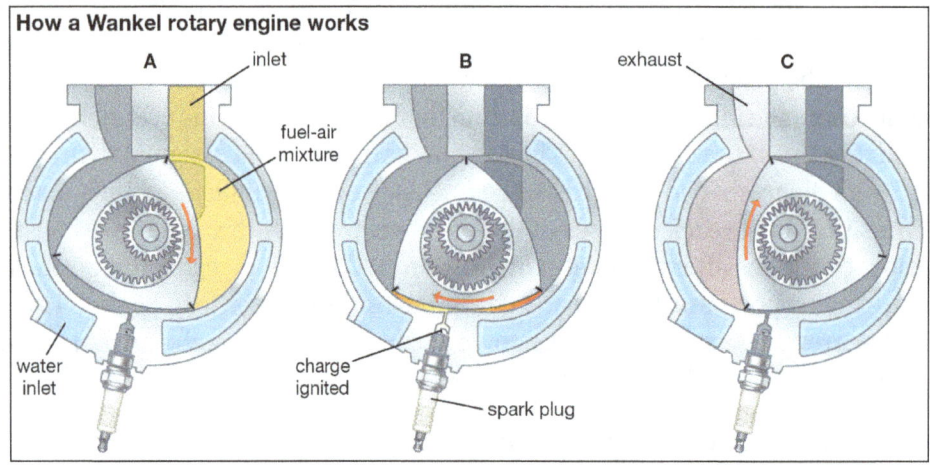

Wankel rotary engine One cycle of the Wankel rotary engine, showing
(A) intake, (B) ignition, and (C) exhaust stages.

In turning about its central axis, the rotor must follow a circular orbit about the geometric centre of the casing. The necessary orbiting rotation is attained by means of a central bore in the rotor in which an internal gear is fitted to mesh with a stationary pinion fixed immovably to the centre of the casing. The rotor is guided by fitting its central bore to an eccentric formed on the output shaft that passes through the centre of the stationary pinion. This eccentric also harnesses the rotor to the shaft so that torque is applied when gas pressure is exerted against the rotor flanks as the fuel and air charges burn. A 3-to-1 gear ratio causes the output shaft to turn three times as fast as the rotor turns about the eccentric. Each quarter turn of the rotor completes an expansion or a compression, permitting intake, compression, expansion, and exhaust to be accomplished during one turn of the rotor. The only moving parts are the rotor and the output shaft.

The fuel mixture is supplied by a carburetor and enters the combustion chambers through an intake port in one of the end plates of the casing. An exhaust port is formed in one of the flattened sides of the casing wall, and a spark plug is located in a pocket communicating with the chambers through a small throat in the opposite side of the casing wall.

The rotor and its gears and bearings are lubricated and cooled by oil circulating through the hollow rotor. The apex vanes are lubricated by a small amount of oil added to the fuel in proportions as low as 1 to 200. Water is circulated through cooling jackets in the casing, the entrance to which is located adjacent to the spark plug, where the temperature tends to be highest.

Maintaining pressure-tight joints by suitable seals at the apexes and on the end faces of the rotor is a major design problem. Radial sliding vanes are fitted in slots at the three apex edges and kept in contact with the casing by expander springs. The end faces of the rotor are sealed by arc-shaped segmental rings fitted in grooves close to the curved edges of the rotor and pressed against the casing by flat springs.

The major advantages of the Wankel engine are its small space requirements and low weight per horsepower, smooth and vibrationless operation, quiet operation, and low manufacturing costs resulting from mechanical simplicity. The absence of inertial forces from reciprocating parts and the elimination of spring-closed poppet valves permit operation at much higher speed than is practical for reciprocating piston engines, an advantage because shaft speed must be high for optimum performance. The induction of fresh fuel mixture and exhaust are more effective because the ports are opened and closed more rapidly than with poppet valves, and gas flow through them is almost continuous. Heat transfer and the resulting cooling requirement are low because the jacketed surface is small. Lower weight and a lower centre of gravity make it much safer in an automobile in the event of a collision. However, competitive fuel economies and the higher development and manufacturing costs of meeting emission standards have limited the use of the Wankel engine in production vehicles, with only the Mazda Motor Corporation marketing any substantial number.

Fuel

Gasoline was originally considered dangerous and was discarded and destroyed at early refineries, which were manufacturing kerosene for lamps. As the gasoline engine developed, gasoline and the engine were harmonized to attain the best possible matching of characteristics. The most important properties of gasoline are its volatility and antiknock quality. Volatility is a measure of the ease of vaporization of gasoline, which is adjusted in the production process to account for seasonal and altitude variations in the local market. Properly formulated gasoline helps engines to start in cold weather and to avoid vapour lock in hot weather.

To suit the needs of a modern engine, a gasoline must have the volatility for which the fuel system of the engine was designed and an antiknock quality sufficient to avoid knock under normal operation. Although other specifications must also be met, volatility and knock rating are the most important. The size and structural arrangement of the molecules principally determine the knocking tendency of a gasoline as well as its volatility.

Tetraethyl lead, added to gasolines for many years to improve antiknock fueling, has been found to contaminate the exhaust gases with poisonous lead oxides, and so the practice has ended. Lower compression ratios and improved combustion-chamber designs have eliminated the need for extremely high-antiknock gasolines.

Lubricating oil is added to gasoline used in crankcase-compression two-stroke-cycle engines.

Performance

The performance of an engine is expressed in terms of power, speed, and fuel economy. The three quantities are evaluated with a dynamometer, a laboratory device that applies a controllable load in the form of resistance to the turning of the crankshaft and also measures the torque exerted at the shaft coupling. The resistance imposed by a dynamometer may be adjusted so that the desired engine speed is established at any throttle position. It is thus possible to run the engine at various speeds throughout its operating range, to continuously maintain these operating conditions, and to measure the precise load and speed at which each run is made. Additional test equipment permits measurement of the exact quantity of fuel consumed, as well as the duration of the runs. From these data the power-speed-economy relationships can be calculated and performance plotted.

The power produced by an engine is expressed in horsepower. When the power developed is measured by means of a dynamometer or similar braking device, it is called brake horsepower. This is the power actually delivered by the engine and is therefore the capacity of the engine. The power developed in the combustion chambers of the engine is greater than the delivered power because of friction and other mechanical losses. This power loss, called the friction horsepower, can be evaluated by "motoring" the engine (driving it in a forward direction) with a suitable dynamometer when no fuel is being burned. The power developed in the cylinder can then be found by adding the friction horsepower to the brake horsepower. This quantity is the indicated horsepower of the engine, so called from an instrument known as the engine indicator, which is used to measure the pressure on the piston and thus calculate the power developed in the cylinder.

Mechanical efficiency is defined as brake horsepower in percent of indicated horsepower and is usually between 70 and 90 percent for normal operating speeds.

A quantity called brake mean effective pressure is obtained by multiplying the mean effective pressure of an engine by its mechanical efficiency. This is a commonly used index expressing the ability of the engine, per unit of cylinder bore, to develop both useful pressure in the cylinders and delivery power. If the power delivered is increased by any change other than an increase in speed or cylinder dimensions, its brake mean effective pressure increases proportionately.

Comparison with other Engines

When the gasoline engine is compared with other types of internal-combustion engines, certain similarities and differences, as well as some advantages and disadvantages, become apparent. The diesel engine and the gas engine (an engine utilizing a gas such as compressed natural gas or propane as the fuel) have a good deal in common with the gasoline engine, since they are all cylinder-and-piston engines that burn air-fuel mixtures in contact with moving components. The important difference that distinguishes the diesel engine is that it has no spark-ignition system. Compared with a gasoline engine of the same horsepower, the diesel engine is heavier and more expensive, but it has a longer life and operates at less cost per horsepower-hour because it burns less fuel.

The gas engine has much in common with the gasoline engine; in fact, in some instances their differences are very slight at best. Structurally, the difference lies primarily in the substitution of a gas-mixing valve for a carburetor. The cylinder and piston configurations are the same. In general, gases have better antiknock qualities than gasoline, permitting slightly higher compression ratios without knock or other combustion difficulties.

From the standpoint of application, the gas engine burning natural gas, manufactured gas, or industrial by-product gas is limited primarily to stationary power plant use because it must remain connected to the gas pipeline. If, however, the fuel is liquefied petroleum gas, sometimes called bottled gas, the containers of gas can be carried in a vehicle, leading to much flexibility in applications. The present obstacle is that facilities are not readily available for replenishing the gas supply.

References

- Amrelweekil (14 September 2009). "Engine modify by Somender Singh". Youtube. Grooved flathead at 1:31–1:38. Retrieved 9 April 2014

- What-are-main-parts-of-automobile-engine: mechanicalbooster.com, Retrieved 16 March, 2019

- Clew, Jeff (2004). The Scott Motorcycle: The Yowling Two-Stroke. Haynes Publishing. P. 240. ISBN 0854291644

- Why-the-boxer-engine, car-advice: subaru.com.au, Retrieved 17 April, 2019

- Ritti, Tullia; Grewe, Klaus; Kessener, Paul (2007), "A Relief of a Water-powered Stone Saw Mill on a Sarcophagus at Hierapolis and its Implications", Journal of Roman Archaeology, 20: 138–163, doi:10.1017/S1047759400005341

- Hunt, Phil; mckay, Malcolm; Wilson, Hugo; Robinson, James (2012), Duckworth, Mick (ed.), Motorcycle: The Definitive Visual History, DK Publishing, Penguin Group, pp. 126, 210, ISBN 9781465400888

- Gasoline-engine, technology: britannica.com, Retrieved 18 May, 2019

- "EPA, California Notify Volkswagen of Clean Air Act Violations / Carmaker allegedly used software that circumvents emissions testing for certain air pollutants". US: EPA. September 18, 2015. Retrieved July 1, 2016

Automotive Parts

Automotive parts include body, automotive valvetrain, suspension system, rolling chassis, backbone chassis, bumper and trunk in a car, automobile wheels, etc. This chapter closely examines these different components of automobiles to provide an extensive understanding of the subject.

BODY

Automotive body designs are frequently categorized according to the number of doors, the arrangement of seats, and the roof structure. Automobile roofs are conventionally supported by pillars on each side of the body. Convertible models with retractable fabric tops rely on the pillar at the side of the windshield for upper body strength, as convertible mechanisms and glass areas are essentially nonstructural. Glass areas have been increased for improved visibility and for aesthetic reasons.

The Fiat 600, introduced in 1956, was an inexpensive, practical car with simple, elegant styling that instantly made it an icon of postwar Italy. Its rear-mounted transverse engine produced sufficient power and saved enough space to allow the passenger compartment to accommodate four people easily.

The high cost of new factory tools makes it impractical for manufacturers to produce totally new designs every year. Completely new designs usually have been programmed on three- to six-year cycles with generally minor refinements appearing during the cycle. In the past, as many as four years of planning and new tool purchasing were needed for a completely new design. Computer-aided design (CAD), testing by use of computer simulations, and computer-aided manufacturing (CAM) techniques may now be used to reduce this time requirement by 50 percent or more.

An automobile being manufactured on an assembly line.

Automotive bodies are generally formed out of sheet steel. The steel is alloyed with various elements to improve its ability to be formed into deeper depressions without wrinkling or tearing in manufacturing presses. Steel is used because of its general availability, low cost, and good workability. For certain applications, however, other materials, such as aluminum, fibreglass, and carbon-fibre reinforced plastic, are used because of their special properties. Polyamide, polyester, polystyrene, polypropylene, and ethylene plastics have been formulated for greater toughness, dent resistance, and resistance to brittle deformation. These materials are used for body panels. Tooling for plastic components generally costs less and requires less time to develop than that for steel components and therefore may be changed by designers at a lower cost.

To protect bodies from corrosive elements and to maintain their strength and appearance, special priming and painting processes are used. Bodies are first dipped in cleaning baths to remove oil and other foreign matter. They then go through a succession of dip and spray cycles. Enamel and acrylic lacquer are both in common use. Electrodeposition of the sprayed paint, a process in which the paint spray is given an electrostatic charge and then attracted to the surface by a high voltage, helps assure that an even coat is applied and that hard-to-reach areas are covered. Ovens with conveyor lines are used to speed the drying process in the factory. Galvanized steel with a protective zinc coating and corrosion-resistant stainless steel are used in body areas that are more likely to corrode.

AUTOMOTIVE VALVETRAIN

The valve train refers to the assembly of components designed to open and close the intake and exhaust valves. Most new engines have overhead cam assemblies like the one shown. Other designs locate the camshaft lower in the engine and use push rods to move valve assemblies. The camshaft is rotated by a timing belt, timing chain or direct gear.

Camshaft

The camshaft is manufactured with precisely machined lobes which control valve opening. The number of lobes on a shaft is determined by the number of valves the shaft controls. Some engines use one shaft to control both intake and exhaust valves. Others have dedicated camshafts for each valve type. Engines designed with four valves per cylinder are normally equipped with dual camshafts for each row of cylinders.

Cam Lobe

Cam lobes are precisely machined into shapes which determine when the valve opens in relation to piston position, how far the valve is displaced, and the length of time the valve remains open. The distance between the end point of the base radius and the nose controls valve displacement. The geometry of the sides (flank) and nose determine how long the valve remains open.

Cam Follower

The cam follower is seated on top of the valve stem and spring and is the surface upon which the cam lobe pushes to open the valve. The follower slides up and down within a bore machined in the cylinder head.

Camshaft Lift

Camshaft lift is the distance between the end point of the lobe's base radius and the nose. The lift determines how far the valve will be displaced. Increasing the lift increases valve displacement.

Camshaft Duration

Camshaft duration is the length of time the valve remains open. The geometry of the lobe's nose and flank determines the duration. A steeply angled flank results in a sharper nose. This produces a shorter duration.

Push Rod

Engines designed with the camshaft located in the engine block use push rods, acting on rocker arms, to open valves. Push rods are seated on valve lifters or tappets which ride on the camshaft lobes. Three types of lifters are used: hydraulic valve lifter, mechanical lifter, and roller lifter. Some push rods are hollow, providing a means to feed oil from the lifters to the rocker arms. This reduces wear on the push rod tip and rocker arm.

Hydraulic Lifters

Hydraulic lifters are used most often since they can reduce valve train noise by maintaining zero valve clearance (no spacing between valve train components). The oil filled lifters adjust automatically for changes brought on by temperature variations and wear of parts. Engine motor oil fills the inside of the lifter, pushing the lifter plunger up until all the play in the valve train is removed.

Mechanical Lifters

Mechanical lifters, also called solid lifters, simply transfer cam lobe action to the push rod. They do not contain oil and are not self adjusting. As a result, they require periodic adjusting. Valve trains using mechanical lifters are prone to a clicking or clattering noise as the valves open and close. This is why hydraulic lifters are more common.

Roller Lifters

Roller lifters are either mechanical or hydraulic. Designed into the lifter is a roller that rides the cam lobe, reducing friction between the camshaft and lifter. Friction between these two components is one of the highest friction points in an engine.

Spring Retainer

The spring retainer is designed to hold in place the valve stem tip. This allows the rocker arm to act directly on the valve.

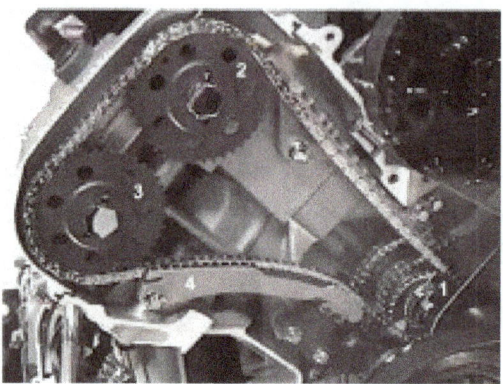

Timing Chain

Timing chains are becoming the standard for turning both the intake and exhaust camshafts. The chains are located on the engine's front end, and are pulled by a drive sprocket, which is turned by the engine's crankshaft. Timing chains are required for both the intake cam sprocket and the exhaust cam sprocket. A chain guide is also provided. Some engines still use belts instead of chains. In either case, excess slack or play will degrade engine performance.

Timing Belt

A timing belt instead of a timing chain may be used to turn the camshafts. The inner side of the belt is designed with square (cogged) teeth which prevent the belt from slipping. The belt should be checked periodically for wear and proper tension.

Belt Tensioner

The belt tensioner is a spring-loaded wheel which keeps the timing belt in tension and aligned with the cam sprocket. The smooth side of the timing belt rides over the tensioner. The tensioner applies a force on the backside of the belt. This keeps the belt in tension. Whenever the belt needs to be removed, the tensioner can be pulled away, freeing the belt.

Valves

Each cylinder has at least one intake valve and one exhaust valve. Some engines are designed with two sets of valves per cylinder as shown in the photo. The intake valve has a larger diameter than the exhaust valve, which maximizes air flow to the cylinder. The exhaust valve must withstand higher temperatures than the intake valve since the air flowing past the intake valve keeps the intake valve at a lower temperature. However, both intake and exhaust valves must transfer their heat to the cylinder head otherwise they will burn.

Valve Springs

Valve springs provide a resisting force that returns displaced valves to their closed position. The spring may be a one-coil design or a two-coil design which has inner and outer coils. The second coil increases the amount of force holding the valve closed.

Sodium Filled Valves

Sodium filled valves are used when extra cooling is required. The hollow valves contain sodium which melts during engine operation. Valve action causes the sodium to circulate, removing heat from the valve head. The heat travels up the valve stem and is transferred to the cylinder head. Coolant channels in the cylinder head (shown in green) carry the heat away.

Stellite Valves

Stellite valves have a hard metal coating that prolongs valve life. Back in the days of leaded gasoline, the lead additives coated the valves, providing added protection. With unleaded fuels now being burned, the hard metal coating does the protecting.

CHASSIS

A chassis is the load-bearing framework of an artificial object, which structurally supports the object in its construction and function. An example of a chassis is a vehicle frame, the underpart of a motor vehicle, on which the body is mounted; if the running gear such as wheels and transmission, and sometimes even the driver's seat, are included, then the assembly is described as a rolling chassis.

Examples of Use

Vehicles

In the case of vehicles, the term rolling chassis means the frame plus the "running gear" like engine, transmission, drive shaft, differential and suspension. An underbody (sometimes referred to as "coachwork"), which is usually not necessary for integrity of the structure, is built on the chassis to complete the vehicle.

For commercial vehicles, a rolling chassis consists of an assembly of all the essential parts of a truck (without the body) to be ready for operation on the road. Car chassis will be different from one for commercial vehicles because of the heavier loads and constant work use. Commercial vehicle manufacturers sell "chassis only", "cowl and chassis", as well as "chassis cab" versions that can be outfitted with specialized bodies. These include motor homes, fire engines, ambulances, box trucks, etc.

In particular applications, such as school buses, a government agency like National Highway Traffic Safety Administration (NHTSA) in the U.S. defines the design standards of chassis and body conversions.

An armoured fighting vehicle's hull serves as the chassis and comprises the bottom part of the AFV that includes the tracks, engine, driver's seat, and crew compartment. This describes the lower hull, although common usage might include the upper hull to mean the AFV without the turret. The hull serves as a basis for platforms on tanks, armoured personnel carriers, combat engineering vehicles, etc.

Electronics

In an electronic device (such as a computer), the chassis consists of a frame or other internal supporting structure on which the circuit boards and other electronics are mounted.

In some designs, such as older sets, the chassis is mounted inside a heavy, rigid cabinet, while in other designs such as modern computer cases, lightweight covers or panels are attached to the chassis.

The combination of chassis and outer covering is sometimes called an enclosure.

Firearms

In firearms, the chassis is a bedding frame on long guns such as rifles to replace the traditionally wooden stock, for the purpose of better accurizing the gun. The chassis is usually made from hard metallic material such as aluminium alloy (and less frequently stainless steel or titanium alloy) due to metals having more superior stiffness and compressive strength compared with wood or synthetic polymer, which are commonly used in conventional rifle stocks.

The chassis essentially functions as a more extensive pillar bedding, providing a metal-on-metal bearing surface that has reduced shifting potential under the stress of recoil. A barreled action bedded into a metal chassis would theoretically operate more consistently during repeated firing, resulting in better precision. With the increasing availability of CNC machining, chassis have become more affordable and sophisticated, and gained increasing popularity as it can also be expanded to accommodate customizable "furnitures" (buttstock, pistol grip, etc.) and rail interface systems that provide mounting points for various accessories.

Rolling Chassis

A rolling chassis is the chassis without bodywork of a self-propelled motor vehicle; car or truck, bus or other commercial vehicle, assembled with its final engine and drivetrain, able to move under its own power.

Heavy Vehicles

2009 Scania rolling chassis heading for bus body maker Lahden Autokori.

1932 rolling chassis for Ford vans.

Separate chassis remain in use for almost all heavy vehicles ranging from pick up trucks to the biggest trucks and commercial passenger carrying vehicles.

The rolling chassis is delivered to the commercial body maker/coachbuilder or bulk transporter on its own wheels under its own power.

Automobiles

Rolling chassis was a stage of manufacture of every vehicle until automobile construction methods changed when Unibody or monocoque combined chassis and body structures gradually replaced chassis.

Restoration

In restoration circles rolling chassis is a name given to a bodyshell with wheels and suspension but without engine and transmission.

Backbone Chassis

Backbone tube chassis is a type of automobile construction chassis that is similar to the body-on-frame design. Instead of a two-dimensional ladder-type structure, it consists of a strong tubular backbone (usually rectangular in cross section) that connects the front and rear suspension attachment areas. A body is then placed on this structure. It was first used in the English Rover 8hp of 1904 and then the French Simplicia automobile in 1909.

The backbone chassis was extensively developed by Hans Ledwinka who used it in greater numbers on the Tatra 11 and subsequent vehicles. Ledwinka later used backbone frame with central tube and axles with swinging driveshafts on Tatra trucks, became known as Tatra-concept.

Design

The chassis of a Škoda Popular (1934).

The truck backbone chassis is a design feature of Czech Tatra heavy trucks (cross-country, military etc.). Hans Ledwinka used this style of chassis for the Tatra 11 car in 1923. He then developed the design on trucks with 6x4 model Tatra 26, which had excellent off-road abilities.

Lotus Elan chassis with rear Chapman strut suspension.

This type of chassis has been used in numerous sports cars. It does not provide protection against side collisions, and thus has to be combined with a body that would compensate for this shortcoming.

Examples of cars using a backbone chassis are Simplicia, De Tomaso Mangusta, DMC De-Lorean, Lloyd 600, Lotus Elan, Lotus Esprit and Europa, Škoda Popular, Škoda Rapid, Škoda Superb, Tatra 77, Tatra 87, Tatra 97 etc. and TVR S1. Trucks with a backbone chassis include the Tatra 111, Tatra 148 and Tatra 815. Some cars also use a backbone as a part of the chassis to strengthen it. Examples include the Volkswagen Beetle, where the transmission tunnel forms a backbone.

De Tomaso P70 racer backbone chassis and running gear. The same parts were later used in production De Tomaso Mangusta.

Hybrid Backbone-ladder Chassis

The Locost may appear to be using a backbone in addition to the outer space frame. But examination shows that, in standard form, it is adding negligible stiffness and only serves as a convenient support structure for the sheet metal panels forming the transmission tunnel. The Triumph Herald and Triumph Vitesse used a twin flanged box section backbone carrying the main torsional and bending loads, with light channel section side rails to stiffen the body, while the Triumph Spitfire and Triumph GT6 sports cars used only the twin-box section backbone, with separate side members in the body, and rear suspension fore and aft loads were also taken by the floor, not the backbone chassis directly.

Advantages

- A standard-conception truck's superstructure has to withstand the torsion twist, and subsequent wear reduces vehicle's lifespan.

- The half-axles have better contact with the ground when operated off-road. This has little importance on roads.

- The vulnerable parts of the drive shaft are covered by a thick tube. The whole system is extremely reliable. However, if a problem occurs, repairs are more complicated.

- The modular system enables configurations of 2-, 3-, 4-, 5-, 6-, or 8-axle vehicles with various wheel bases.

Disadvantages

- Manufacturing the backbone chassis is more complicated and more costly. However, the more axles with all-wheel drive are needed, the cost benefit turns in favor of backbone chassis.

- The backbone chassis is heavier for a given torsional stiffness than a uni-body.

- The chassis gives no protection against side impacts.

BUMPER

A bumper is a structure attached to or integrated with the front and rear ends of a motor vehicle, to absorb impact in a minor collision, ideally minimizing repair costs. Stiff metal bumpers appeared on automobiles as early as 1904 that had a mainly ornamental function. Numerous developments, improvements in materials and technologies, as well as greater focus on functionality for protecting vehicle components and improving safety have changed bumpers over the years. Bumpers ideally minimize height mismatches between vehicles and protect pedestrians from injury. Regulatory measures have been enacted to reduce vehicle repair costs and, more recently, impact on pedestrians.

Physics

Bumpers offer protection to other vehicle components by dissipating the kinetic energy generated by an impact. This energy is a function of vehicle mass and velocity squared. The kinetic energy is equal to 1/2 the product of the mass and the square of the speed. In formula form:

$$E_k = \frac{1}{2}mv^2$$

A bumper that protects vehicle components from damage at 5 miles per hour must be four times stronger than a bumper that protects at 2.5 miles per hour, with the collision energy dissipation concentrated at the extreme front and rear of the vehicle. Small increases in bumper protection can lead to weight gain and loss of fuel efficiency.

Until 1959, such rigidity was seen as beneficial to occupant safety among automotive engineers. Modern theories of vehicle crashworthiness point in the opposite direction, towards vehicles that crumple progressively. A completely rigid vehicle might have excellent bumper protection for vehicle components, but would offer poor occupant safety.

Pedestrian Safety

Bumpers are increasingly being designed to mitigate injury to pedestrians struck by cars, such as through the use of bumper covers made of flexible materials. Front bumpers, especially, have been lowered and made of softer materials, such as foams and crushable plastics, to reduce the severity of impact on legs.

Height Mismatches

For passenger cars, the height and placement of bumpers is legally specified under both US and EU regulations. Bumpers do not protect against moderate speed collisions, because during emergency braking, suspension changes the pitch of each vehicle, so bumpers can bypass each other when the vehicles collide. Preventing override and underride can be accomplished by extremely tall bumper surfaces. Active suspension is another solution to keeping the vehicle level.

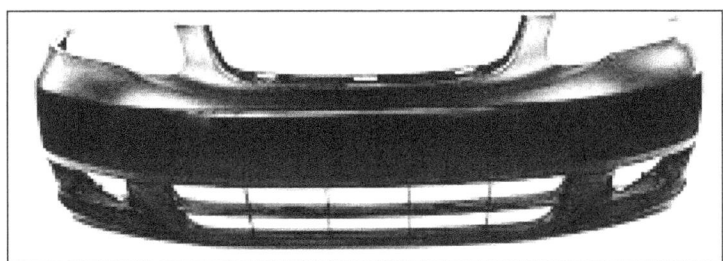

Bumper height from the roadway surface is important in engaging other protective systems. Airbag deployment sensors typically do not trigger until contact with an obstruction, and it is important that front bumpers be the first parts of a vehicle to make contact in the event of a frontal collision, to leave sufficient time to inflate the protective cushions.

Energy-absorbing crush zones are completely ineffective if they are physically bypassed; an extreme example of this occurs when the elevated platform of a tractor-trailer completely misses the front bumper of a passenger car, and first contact is with the glass windshield of the passenger compartment.

Truck vs. Car

Underride collisions, in which a smaller vehicle such as a passenger sedan slides under a larger vehicle such as a tractor-trailer often result in severe injuries or fatalities. The platform bed of a typical tractor-trailer is at the head height of seated adults in a typical passenger car, and can cause severe head trauma in even a moderate-speed collision. Around 500 people are killed this way in the United States annually.

Following the 1967 death of actress Jayne Mansfield in an auto/truck accident, the US government agency NHTSA recommended requiring a rear underride guard, also known as a "Mansfield bar", an "ICC bar", or a "DOT (Department of Transportation) bumper". They are required to be not more than 22 in (56 cm) from the road. The trucking industry has been slow to upgrade this safety feature, and there are no requirements to repair ICC bars damaged in service. However, in 1996 NHTSA upgraded the requirements for the rear underride prevention structure on truck trailers, and Transport Canada went further with an even more stringent requirement for energy-absorbing rear underride guards, and in July 2015 NHTSA issued a proposal to upgrade the US performance requirements for underride guards.

Many European nations have also required side underride guards, to mitigate against lethal collisions where the car impacts the truck from the side. A variety of different types of side underride guards of this nature are in use in Japan, the US, and Canada. However, they are not required in the United States.

SUV vs. Car

Modest mismatches between SUV bumper heights and passenger car side door protection have allowed serious injuries at relatively low speeds. Unlike trucks, SUVs with bumpers more than 22 in (56 cm) from the road are illegal in the United States, as are vehicles with the fuel tank located behind the rear axle. In the United States, NHTSA is studying how to address this issue as of 2014.

Beyond lethal interactions, repair costs of passenger car/SUV collisions can also be significant due to the height mismatch. This mismatch can result in vehicles being so severely damaged that they are inoperable after low speed collisions.

TRUNK

The trunk of a car is the vehicle's main storage or cargo compartment.

Designs

The trunk or luggage compartment is most often located at the rear of the vehicle. Early designs included an exterior rack mounted on the rear of the vehicle to which it was possible to attach a real luggage trunk. Later designs integrated the storage area into the vehicle's body and evolved to provide a streamlined appearance. The main storage compartment is normally provided at the end of the vehicle opposite to which the engine is located.

Some mid-engined or electric cars have luggage compartments both in the front and in the rear of the vehicle. Examples include the Volkswagen Type 3, Porsche 914, Porsche Boxster, Toyota MR2, and Tesla Model S (Tesla calls their front trunk a "frunk"). The mid-engined Fiat X1/9 also has two storage compartments, although the rear one is small, easily accessible, and practically cuboid in shape.

Rear-engined cars (such as the Volkswagen Beetle, Tucker 48, and the Porsche 911) have the trunk situated in front of the passenger compartment.

Sometimes during the design life of the vehicle the lid may be restyled to increase the size or improve the practicality and usefulness of the trunk's shape. Examples of this include the Beetle redesign to the 1970s 'Super Beetle' and the pre-war and 1950s post war Citroën Traction Avant.

Openings

Door

The door or opening of a cargo area may be hinged at the top, side, or bottom.

If the door is hinged at the bottom it is termed a tailgate, particularly in the United States. They are used on station wagons and pickup trucks, as well as on some sport utility vehicles (SUV). Traditional drop-down station wagon and pickup tailgates can also serve as a mount for a workbench.

Traditional U.S. station wagons included a roll down window retracting into the tailgate to load small items or to allow the tailgate to be opened down on its bottom mounted hinges. Because of the potential for carbon-monoxide fumes, the tailgate window on station wagons should be closed whenever the engine is running.

Two-way station wagon tailgates may be hinged at the side and the bottom so they can be opened sideways like a regular door, or drop downwards as load platform extenders. They are designed with special handle(s) for opening in the selected direction on special hinges after the window is lowered.

A three-way design that was also used by Ford allows for the tailgate to be opened like a door with the window up.

General Motors developed a clam shell style "disappearing" design where the rear window rolls up into the roof and the tailgate slides down and beneath the load floor.

If the door is hinged at the top it is termed a hatch, and is used on a hatchback. A bottom opening door is now common on sport utility vehicles (SUV).

Lid

The trunk lid (in the U.S. automotive industry sometimes also called decklid or deck lid) is the cover that allows access to the main storage or luggage compartment. Hinges allow the lid to be raised. Devices such as a manually positioned prop rod can keep the panel up in the open position. Counterbalancing torsion or other spring(s) can also be used to help elevate and hold open the trunk lid. On cars with their trunk in the rear, lids sometimes incorporate a center mounted third brake light. A rear lid may also have a decorative air spoiler. On many modern cars, the trunk lids can be unlocked with the car's key fob.

Locks

The locking of the trunk may be achieved together with the passenger compartment.

Some cars include a function to remotely open the trunk. This may be achieved through a variety of means:

- Release of the latch whereby the doorseals push the decklid away from the lock, the trunk is then open, and the lid may not have revealed the opening.

- Release of the latch whereby a spring pushes the decklid away from the lock and open, the trunk is then open, and the lid reveals the opening.

- Release of the latch and actuation of a drive, whether hydraulic (BMW 7 Series) or electric (BMW X6), which pushes the decklid away from the lock; the trunk is then open, and the lid reveals the opening. This may then be electrically closed again.

Classification

Open or Closed Compartments

Open compartments are those found in station wagons and SUVs, while closed compartments have a trunk lid and are typically found in saloon (sedan) or coupé bodies. Closed compartments are separated from the passenger compartment by rigid body elements or seats, and are generally trimmed in simple materials, whereas many station wagons are trimmed with better looking materials as the space is an extension of the passenger compartment. In order to hide the compartment content of station wagons or hatchbacks from thieves or sunlight, a cover may be fitted. On hatchbacks this often has the form of a rigid parcel shelf or a flexible sheet with hooks on the corners, while station wagons and many SUVs have a roller blind in a removable cassette.

Increased Variability

To give the space more flexibility, many cars have foldable rear seats, which can increase the size of the trunk when needed.

Safety

Active Safety by Luggage Retention

The trunk space can contribute to the active and passive safety of the vehicle. Active safety may be promoted in vehicles that are partially loaded. Here the use of lashing eyes to restrain luggage can prevent or reduce damage to the vehicle and its occupants in severe manoeuvres. In driving while cornering 'in-extremis', the prevention of sudden weight transfer due to poorly loaded luggage can be enough to prevent the vehicle losing grip, and potentially avoiding thereby an accident; active safety.

Passive Safety by Luggage Retention

If a crash should occur, lashing eyes can reduce the severity of outcome of the accident by keeping the luggage in the loadspace compartment and thereby preventing projectiles from harming correctly restrained passengers in the passenger compartment. These lashing features may be in the form of fixed or foldable loops, or in the case of certain European vehicles combine sliding loops in a rail system to allow optimal positioning of the lashing eyes. At the same time this eases the integration of accessories for loadspace management; dividers, bike carriers, etc. into the interior of the vehicle, a principle that has been applied in cargo vans and air transport for many years.

Barrier Nets/Grids

In vehicles with open luggage compartments, some are fitted with metal grids or guards to retain loose items in case of collision, or to simply create a bulkhead between the load in the trunk – for example animals – separated from the otherwise unprotected passenger space.

Another solution for items that have not been restrained is the loadspace barrier net. These may be directly attached to the body structure or, in vehicles with loadspace cover cassettes, as a combined loadspace cover and barrier net. The net confining luggage to the loadspace in case of emergency braking and minor traffic collisions. These nets have the advantage over metal guards that they can be rolled-up when not in use, taking up much less space than a comparable guard. A guard may however be tailored for an even tighter fit to the body interior contours than a roll-away net.

Inside Trunk Release

Children – and sometimes adults who climb in to work on the vehicle – trapped in trunks can die of suffocation or heat stroke. Once in the trunk, they may not be able to get out, even if they entered through the interior, because many rear seats only release to the trunk from inside the passenger area. Beginning with the 2002 models, a glow-in-the-dark inside trunk release is required on all vehicles with conventional trunks sold in the United States. Hatchbacks, wagons, vans and SUVs are exempt from this requirement because it is assumed a trapped person can kick out any cargo cover or parcel shelf to gain access to the main interior and passenger doors.

Riding in the Trunk

Riding in the trunk is dangerous and illegal. Teenagers in the U.S. may attempt to avoid laws which prohibit new car drivers from driving with passengers by riding in the trunk.

Additional Functions

Beyond carrying luggage, the trunk of most passenger vehicles commonly contains various other components often behind the trimmed surfaces of the interior. These components may be accessed by the customer or the service personnel through (in some cases lockable) hatches in the trim, or by removing carpet and support boards etc. Typical components:

- Emergency supplies.
- Spare tire.
- Jack and lug wrench.
- On-board tool kit for do it yourself repairs.
- Electronics for sound, video, GPS, etc.
- Battery and hybrid energy store.
- Fuse boxes.
- CNG/LPG tanks (for bivalent engines).
- Additional folding, or 'third-row', seating (increasingly in open loadspaces).

Some vehicles offer configurable cargo conveniences such as a shelf or board. They often serve various purposes. The multiposition rear shelf on the Chrysler PT Cruiser can be used as a table for a picnic, a second cargo layer, or a security screen. The Citroën C3 has a foldable segmented false floorboard that compartmentalises the cargo area, makes loading easier, and evens out the load floor when the back of the rear seat is folded down.

AUTOMOBILE WHEELS

The automobile wheels support the total weight, resist the strains created during turning and transmit the driving torque for propelling and breaking torque for retarding. The wheels must have well balanced construction particularly for running at high speeds. Unbalanced wheel assemblies cause excessive vibration, which accelerates tyre and king-pin wear.Automobile control, acceleration and braking occur through the tyres and their contact on the road surface. The tyres must be large and strong enough to support the vehicle on the road. The traction, the force required to make the tyre slip on the contact patch, is the same on the wheel in all directions whether it is accelerating, cornering, braking, or any combination of these. The tractive force to control the automobile drops rapidly when a skid starts so that control is lost. The tyre must absorb, by deflecting, part of the shock from road irregularities. During normal operation, passenger car tyres rotate approximately 500 revolutions for each km travelled. The chief factors affecting tyre life are inflation pressure, vehicle speed and rate of acceleration, temperature, tyre interchanging, and tyre and wheel balance. A tractor tyre normally is expected to last twenty years, a truck tyre 80,500 km running, a passenger car tyre 48,500 km and a racing tyre 800 km.

Automobile Wheels

The automobile wheels fulfill a number of following objectives.

Structure

Wheels must be rigid enough to retain their shape under all operating conditions. When subjected to abnormal impact, they should preferably buckle and must not collapse. The dimensional tolerances of the wheel should be accurate enough for carrying out wheel alignment and balancing. Weight. Wheels must be light so that the unsprung weight is reduced. Light wheels and tyres also follow the road surface contour more accurately so that wheel bounce is minimized, resulting in improved road contact and reduced tyre wear.

Tyre Attachment

Since the wheel-tyre combination is responsible for the transmission of traction to the road or for steering reaction, the tyre must be suitably located and rigidly secured on the wheel. Aditionally, the design of wheel should be such that the tyre can be fitted easily.

Wheel Mounting

The wheel attachment must be designed properly for locating, securing, and supporting the wheel. Also the wheel should be easily fitted or removed from its axle-hub.

Cost

Wheels should be made out of cheaper materials that can easily be fabricated, cast, or forged, with the minimum machining. It should also have better finish and appearance and should not easily deteriorate with age and weathering.

Pressed-steel-disc Wheels

These wheels use a formed disc pressed into a rolled-section well-base rim and held in position by spot welding. The formed disc contacts the rim in a number of equal-spaced arcs. The disc has a number of equal-spaced shallow elongated slots immediately under the base of the rim well, which improves brake cooling and decreases the transfer.

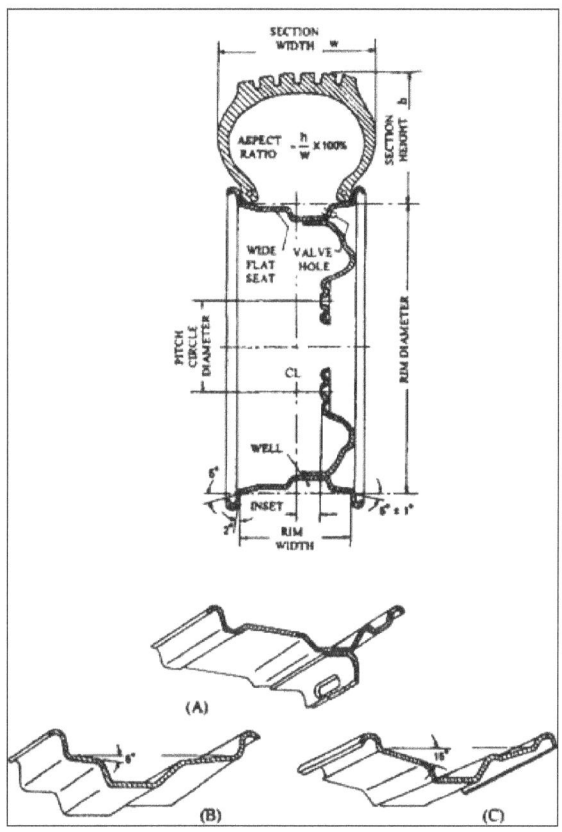

Car well-base disc-type wheel.

Pressed-steel-disc wheel rim A. Car double-hump well-base rim. B. Van 5 degrees seat-angle well-base rim. C. U-type drop centre 15 degrees taper rim. of heat from the brake-drum to the tyre due to passage of air through the slots. The disc is fixed to the axle-hub by a number of studs and nuts with conical or spherical seating. The rim for the car wheel is cold rolled from flat steel strip and the rims for heavier commercial vehicle are hot rolled to the section from steel bars. The steel strip is cut to the required length and then rolled into a circle before the ends are butt-welded. The circular steel strip then undergoes a series of rolling operations to obtain a complete rim. The rim is then expanded to size and the valve hole is punched. The well in the base of the rim is sufficiently deep to hold the beads of the tyre for mounting or demounting. The portion of the rim where the tyre sits have a 5 degrees taper due to which, as the tyre is inflated, the beads are forced up the taper providing a wedge fit, and a good seal is obtained with tubeless tyres.

Car Well-base Rims

Wheel rims are designed to minimize dislodging of the tyre, when subjected to heavy cornering. Originally the rim was tapered from the rim flange to the edge of the well so that the pressure acting over a short length of the bead would force back the tyre bead into the well. This can happen due to the interaction of the cornering side-force acting on the wheel and the reaction of the tyre and the road. This is prevented in the flat-ledge rim profile due to the provision of a parallel flat ledge between the outer taper adj acent to the flange and the edge of the well. In a close-fitting tyre bead this can happen only due to the pressure acting all the way round the bead. Another approach for improving the bead rim seal and joint is to form a hump slightly in front the shorter taper next to the rim flange. Once the bead sits over the short taper portion adjacent to the rim flange, it is very difficult for it to climb over the hump, back into the rim well.

Van 5 Degrees Seat-angle Well-base Rim

These rims have high wall flanges, which protect the tyre beads and walls from damage due to external interference. Generally both tubed and tubeless tyres are fitted on these rims. However large stiff tyres cannot be mounted over these flanges, and hence these rims are only used for vans and small commercial vehicles.

U-type Drop-centre 15 Degrees Taper Rim

Tubeless tyres and large section tyres with relatively flexible beads are used for vans, buses, and medium sized trucks. These tyres use a single-piece well-base wheel rim, having relatively shallow flanges at the edge of 15 degrees taper bead-seats. This profile of the rim provides a good joint and sound seal between the tyre bead and the rim taper.

Detachable-rim Wheels

Large tyres for commercial vehicles use more plies in the casing and are therefore relatively bulkier in the bead region. The rims for these tyres have one removable side-flange, which allows the wheel tyres to slide into position, and then the flange can be replaced and locked in place.

Semi-drop-centre Two-piece Rim

This type of rim has inner and outer tapered tyre-bead-seating surfaces separated by a shallow central or near central well. The inner flange can be removed for fitting a tyre. In this design, the well depth only permits the tyre beads to pass over the top of the outer bead-seat taper. The outer flange is detachable and is sprung into a continuous groove formed along the outer edge of the rim base so that it is retained in its working position. This rim forms an intermediate class between the well-base and wide-base rims. It accommodates heavier beads, which are too rigid for fitting on the full well-base type. This rim is used on light trucks.

Wide-base Two-piece Rim

This rim is made up of one fixed flange integral with the rim base and one split detachable flange. The rim uses 5 degrees taper seats for tyre beads, the one on the detachable flange side being integral with the flange. The outer detachable flange is sprung into a continuous groove formed along the outer edge of the rim base to retain the flange in its working position. During inflation of the tyre the bead sits over the detachable-flange taper, holding it is position. This rim is used mainly on medium-size commercial vehicles.

Wide-base Three-piece Rim

This type of rim contains one fixed flange integral with the rim base, one detachable endless flange, and a separate flange-retaining split lock-ring. The rim uses 5 degrees taper seats for tyre-bead location, the one on the detachable-flange side usually being on an extension of the spring lock-ring. During inflation of the tyre the bead sits over the extension of the spring lock-ring, holding it in position. This rim is used on large commercial vehicles.

Divided Flat-base Rim

This type of rim is integral with the wheel itself. For fitting or removal of the tyre the two halves of the wheel are divided by dismantling the outer ring of bolts, which hold the wheel halves together. These rims are used primarily for large military trucks.

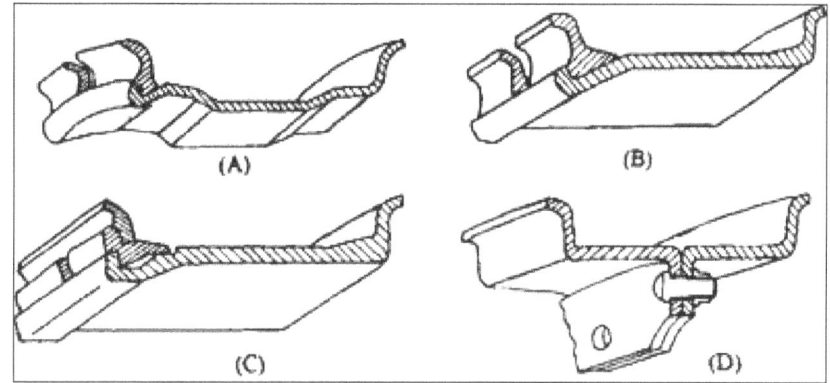

Detachable rims. A. Semi-drop-centre two-piece rim. B. Wide-base two-piece rim.
C. Wide-base Three-piece rim. D. Divided flat-base rim.

Cast and Forged Alloy Wheels

These wheels are manufactured as a single-piece rim and disc. Car wheels are generally cast or extruded, but truck wheels are forged. Magnesium and aluminium alloys are most commonly used for reduction in weight. Magnesium alloy produces a saving of weight of 30% over aluminium alloys and 50% over steel for similar strength. Magnesium alloys exhibit very good fatigue properties and excellent resilience, due to which they are capable of resisting vibrational and shock loading better than both aluminium alloy and steel. However they are highly susceptible to corrosion and therefore must be provided with a protective surface coating. Although aluminium alloys have relatively less fatigue properties but their corrosion characteristic is also less and they can be readily cast or forged. The section thickness for the rim and disc are required to be greater with light alloys than with steel. Even though it is a disadvantage to machine the wheel rim and the stud hole flange after casting or forging operation, but this produces close-tolerance wheels. Also most light alloys are better conductors of heat than steel, so that they transfer any heat generated by the tyre or brake more quickly to the wheel-disc for dissipation to the air stream. Light-alloy wheels are more expensive to manufacture than pressed-steel wheels. Aluminium-alloy wheels are cheaper than magnesium alloy wheels. Light-alloy wheels are used for better appearance and lighter weight. Usually an aluminium alloy is preferred for passenger cars and trucks, and a magnesium alloy for sports and racing cars. The wheel-rim profile used for both light alloy wheels and steel rims is similar.

Wheel Mountings

Conical-nut Mounting

Car steel wheels are generally aligned with conical-taper nuts those fit into the wheel-disc countersink stud holes. The wheel is centralized to the hub axis due to the taper, which also provides a wedge action to the nuts when they are tightened. The wheel then properly located, secured, and held by the stud and nut.

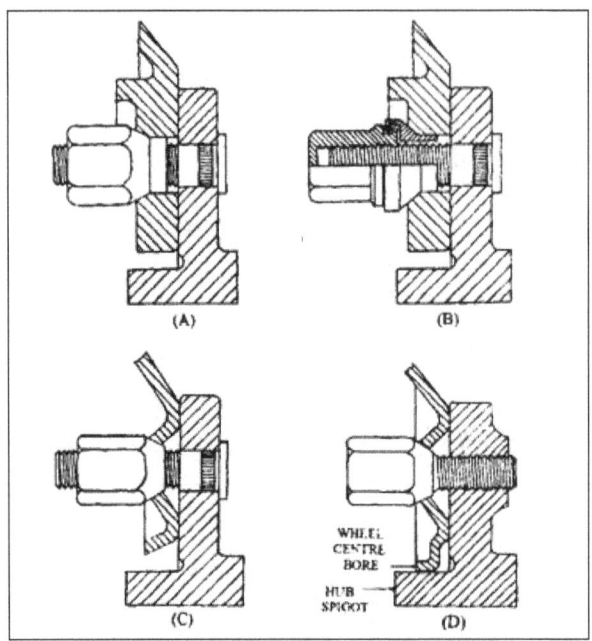

Passenger car wheel fixing. A. Nut with conical seat. B. Spigot-mounting with conical set-bolt.
C. Nut with fitted shoulder and conical seat. D. Nut with fitted sleeve and conical seat.

Spigot Mounting with Conical Set-bolt

Figure represents the spigot-mounting wheel on the axle-hub, which accurately locates and partially supports the wheel. The disc is secured to the hub by the conical-shaped set-bolts, which also transmit both the driving and braking torques. This method is suitable with both steel and light-alloy wheels and is commonly used on small and medium sized cars.

Shouldered-nut Mounting

The light-alloy car wheels with the softer metal may distort or tear away at the stud holes when the nuts are tightened, so that these holes are slowly enlarged. To maintain a concentricity of the wheel, the conical end of the nut is extended with a parallel shoulder or sleeve portion, which is a close fit with the stud hole. The wheel is then located, secured and supported by the nut.

Sleeved-integral-nut Mounting

A further improvement in car wheels is achieved by incorporating a separate sleeve, which has a conical-and-parallel profile. This sleeve rotates freely relative to the hexagon nut and prevents tearing between the alloy wheel and the stud nut during tightening.

Commercial-vehicle Wheels

Conical Taper-nut Mounting

Commercial-vehicle steel wheels are attached to the axle-hub flange by using conical-taper nuts fitting into countersink stud holes. In case of twin wheels, the wheel stud-hole flange next to the drum is located by a loose conical-taper-shaped washer placed over the stud or by an integral spherical-shaped seat on the stud. The wheel is located, secured, and supported by the nut.

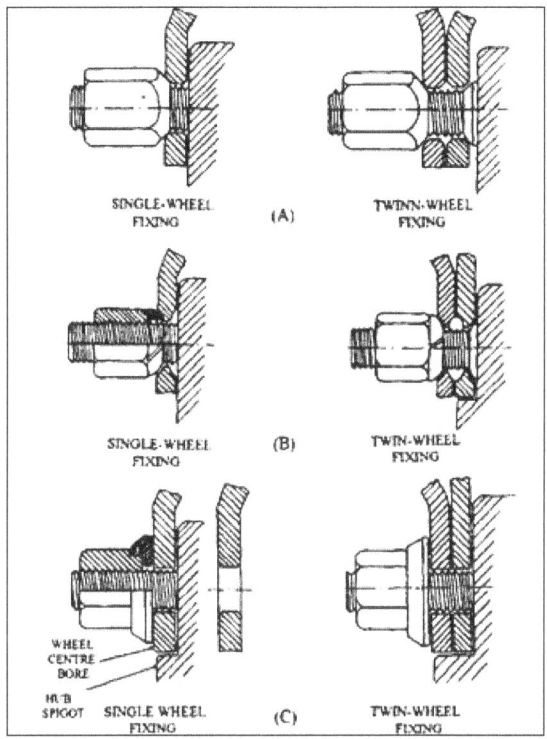

Commercial-vehicle wheel fixings. A. British standard wheel fixing. B. Continental
(DIN) standard wheel fixing. C. Spigot-mounted wheel fixing.

Continental (DIN) Standard Wheel Mounting

The wheel can also be mounted to the hub by plain nuts with split spherical-faced washers. These washers fit into spherical countersinks formed in the wheel stud holes. The spherical seats centralize the stud holes, and the washers and the nuts mainly take the relative motion. The wheel load is supported by the centre-bore of the wheel flange. In case of twin wheels, a spherical washer or seat supports the inner-wheel flange.

Spigot Mounting

With spigot-mounting, the centre-bore has a close-tolerance fit on the hub spigot for locating and supporting the load. The wheel nuts use an integral flat-face washer, which holds the wheel securely to the brake-drum, whereas the stud takes the acceleration and braking torques. The same mounting layout also holds good for twin r«ar wheel arrangements, the support in this case being between the centre-bore pf the wheel stud-hole flange and the hub spigot. This type of wheel mounting is mostly used for large trucks.

References

- Company, Houghton Mifflin Harcourt Publishing. "The American Heritage Dictionary entry: chassis". Www. ahdictionary.com. Retrieved 2017-05-21

- Automobile, technology: britannica.com, Retrieved 19 June, 2019

- Grenzeback, Lance R.; Lin, Sandi; Meunier, Jacob (2005). Operational Differences and Similarities among the Motorcoach, School Bus, and Trucking Industries. Transportation Research Board. P. 13. ISBN 978-0-309-08821-3. Retrieved 10 September 2010

- Valve-train, rm-preview: autoeducation.com, Retrieved 20 July, 2019

- Shuler, S.; Mooijman, F.; Nanda, A. (8 March 2004). "Bumper Systems Designed for Both Pedestrian Protection and FMVSS Requirements: Part Design and Testing". SAE International. Doi:10.4271/2004-01-1610. Retrieved 2 July 2015

- Automobile-wheels-automobile, automobile: what-when-how.com, Retrieved 21 August, 2019

- Grenzeback, Lance R.; Lin, Sandi; Meunier, Jacob (2005). Operational Differences and Similarities among the Motorcoach, School Bus, and Trucking Industries. Transportation Research Board. P. 13. ISBN 978-0-309-08821-3. Retrieved 10 September 2010

Automobile Systems

Automobile systems consist of independent parts that are capable of functioning by itself. Its major system includes engine lubrication system, cooling system, electrical system, transmission system, etc. This chapter delves into these major automobile systems to provide an in-depth understanding of the subject.

ENGINE LUBRICATION SYSTEM

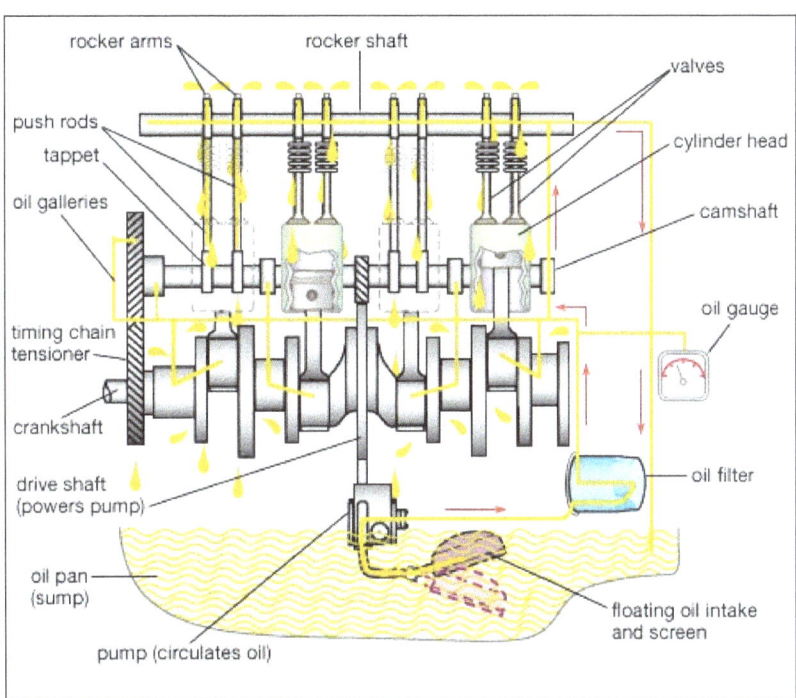

The engine lubrication system is to distribute oil to the moving parts to reduce friction between surfaces. Lubrication plays a key role in the life expectancy of an automotive engine. If the lubricating system fail, an engine would succumb to overheating and seizing very quickly. An oil pump is located on the bottom of the engine. The oil is pulled through a strainer, by the oil pump, removing larger contaminants from the mass of the fluid. The oil then forced through an oil filter under pressure to the main bearings and the oil pressure gauge. It is important to note that not all filters perform the same. A filter's ability to remove particles is dependent upon many factors, including the media material (pore size, surface area and depth of filter), the differential pressure across the media, and the flow rate across the media. From the main bearings, the oil passes into drilled passages in the crankshaft and the big-end bearings of the connecting rod. The oil fling dispersed by the rotating crankshaft lubricates the cylinder walls and piston-pin bearings. The excess oil is scraped off by the scraper rings on the piston. The engine oil also lubricates camshaft bearings and the timing chain or gears on the camshaft drive. The excess oil in the system then drains back to the sump.

Engine Oil

Superior quality engine oil is formulated with the high quality base oil and advanced technology based additive package to provide protection for automotive engines in severe service applications.

SAE Stands for the Society of Automotive Engineers, based in the U.S.A. The SAE grade specifies the most important parameters for engine oil mainly its viscosity. The SAE viscosity classification defines mainly viscosity limits at high and low temperature for any grade of lubricants. The SAE grade guide us to the right viscosity for different outside temperatures. Grades marked 'w' stand for winter are at a temperature below 0 °C.

API stands for the American Petroleum Institute. This body has specified the performance standards that oils used in road vehicles should meet. For oils to use in passenger car engines, the letters API are followed by a set of two letters such as SM, etc. Service Levels for passenger car oils or 'S' indicates for Spark Ignition Engine. These specified performance levels have evolved through the years, from API SA to SN, Similarly, the API designates the performance of diesel engine oils with a letter sequence such as API CF-4.'C' indicates for commercial or compression ignition engine. Automotive gear oils they use API GL-4.API GL-5 etc. The highest API for commercial engine oils (diesel oils) today is API CJ-4.

Working of an Engine Lubrication System

The Engine lubrication system is considered to give a flow to the clean oil at the accurate temperature, with a appropriate pressure to each part of the engine. The oil is sucked out into the pump from the sump, as a heart of the system, than forced between the oil filter and pressure is fed to the main bearings and also to the oil pressure gauge. The oil passes through the main bearings feed-holes into the drilled passages which is in the crankshaft and on to the bearings of the connecting rod. The bearings of the piston-pin and cylinder walls get lubricated oil which dispersed by the rotating crankshaft. By the lower ring in the piston the excess being scraped. Each camshaft bearing is fed by the main supply passage from a branch or tributary. And there is another branch which supplies the gears or timing chain on the drive of camshaft. The oil which is excesses then drains back to the sump, where the heat is being transferred to the surrounding air.

Journal Bearings

If the crankshaft journals get worn, the engine will be having very low oil pressure and will throw oil all over inside the engine. The unnecessary splash will overcome the rings and can cause the engine to use that oil. Simply replacing the bearing inserts can restore the worn bearing surfaces. In well maintained engine, bearing wear take places instantly after a cold start because there is less or no oil film between the shaft and bearing. At the time that enough automotive lubricants is dispersed through the hydrodynamic lubrication system apparent and stops the bearing wear progress.

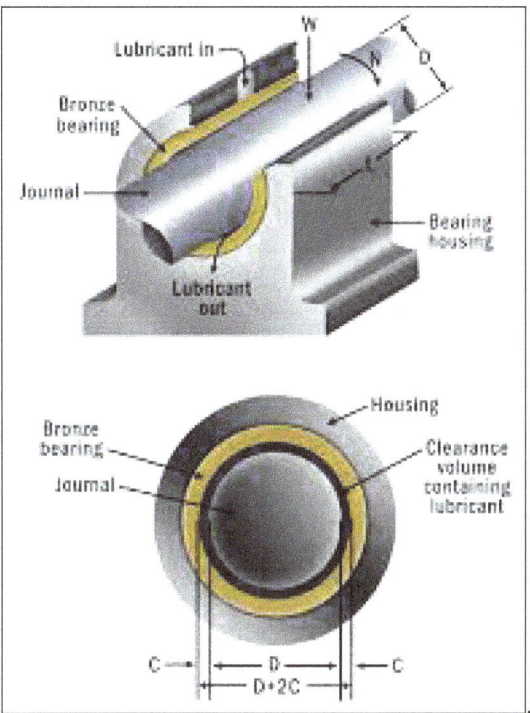

Lubrication Systems for Petrol Engines

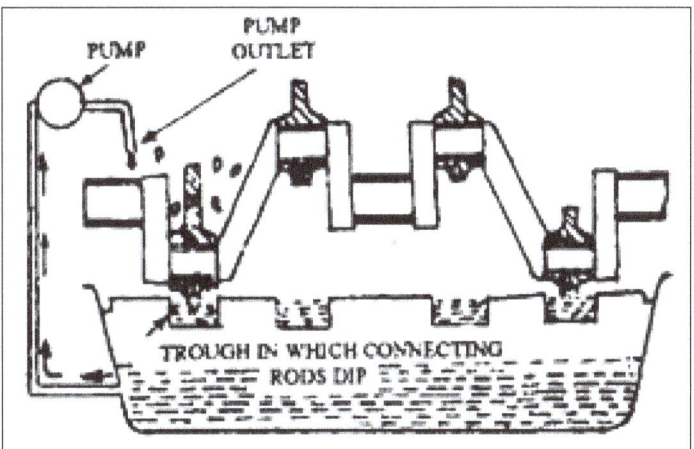

Splash lubrication system.

In order to ensure adequate supplies of oil to the engine parts, a reservoir of oil is provided by the sump which is the lower part of the lubrication system and in automobile engines the sump is the oil pan. From the reservoir, oil is distributed throughout the engine either by the splash system or the full pressure system. In case of two-stroke engines, the crankcase cannot be used as an oil reservoir. The lubrication, in this case, is provided by mixing a small proportion of oil with petrol.

In the splash system the oil is maintained in little troughs. There are dippers at the ends of the connecting rods to splash the oil on the various parts like cylinder walls, camshafts, gudgeon pins etc., as they travel through the oil troughs towards the bottom of the stroke of the piston. The oil is supplied to the main bearings under pressure due to an oil pump through drilled passages, in the crankcase, called galleries. The oil pump also replenishes the troughs. The system is now practically obsolete.

Full Pressure System

Automobile engines today use 'forced-feed' lubrication systems, generally of the wet-sump type in which the sump acts as both an oil-drain return and a storage container. A rotary-type oil-pump provides forced feed. The pump may be driven directly from the crankshaft or indirectly from the camshaft or any auxiliary shaft. Oil from the sump reaches the pump through the submerged gauze strainer and pick-up pipe. The oil is then compressed, which passes through a drilling to the lubrication system. A pressure-relief valve positioned on the output side of the pump controls the oil pressure. If the oil pressure becomes too high, the relief valve opens and bleeds surplus oil back to the sump. The relief valve may be installed on the filter unit, the crankcase, or the pump housing. The oil-pump forces the oil through drillings in the crankcase to a cylindrical full-flow filter unit. The oil circulates around the filter bowl, passes through the filter towards its centre, and flows out to the main oil passage, called main oil gallery which lies parallel to the crankshaft. In most car and commercial vehicle engines, the oil gallery is formed by drilling a hole in the crankcase for full length of the engine and plugging the ends.

Main- and Big-end Bearing Lubrication

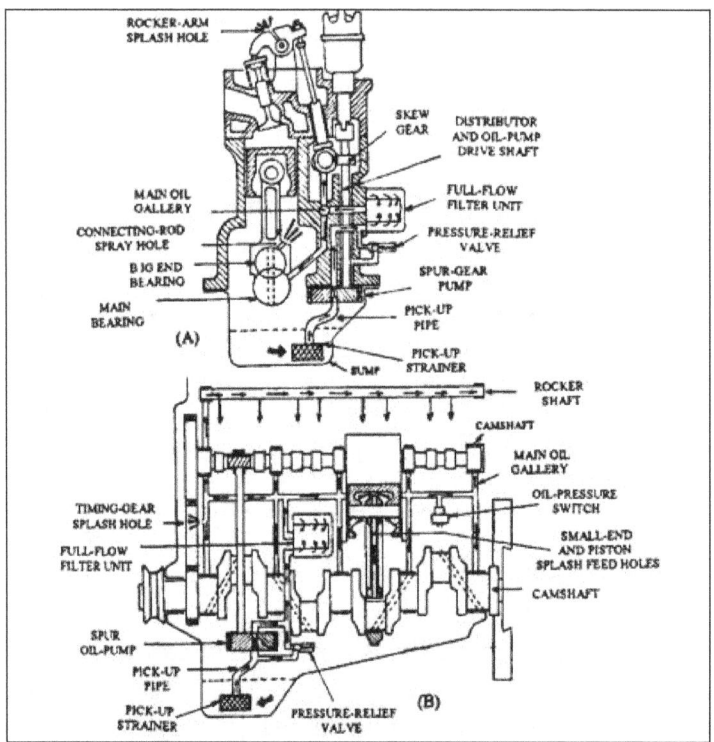

Forced-feed lubrication system. A. Front sectional view. B. Side sectional view.

The oil is fed to the crankshaft main journal bearings and in some cases to the camshaft bearings through various branch cross-drillings in the crankcase. A few heavy commercial engines use a separate pipe located underneath the main-bearing caps and by pedestal brackets. Drillings in these brackets connect the gallery-pipe oil to the main bearings. By diagonal drillings in the crank-shaft a continuous oil is fed to the big-end bearings from the oil grooves around the main-bearings liners. These drillings pass from the main-bearing journal to the big-end crankpins through the crankshaft web.

Cylinder and Piston Lubrication

Four separate techniques are used for cylinder and piston lubrication.

- Connecting-rod big-end side-clearance oil spray.

- Connecting-rod big-end radial-hole oil spray.

- Connecting-rod small-end radial-hole oil spray.

- Crankcase fixed-jet oil spray.

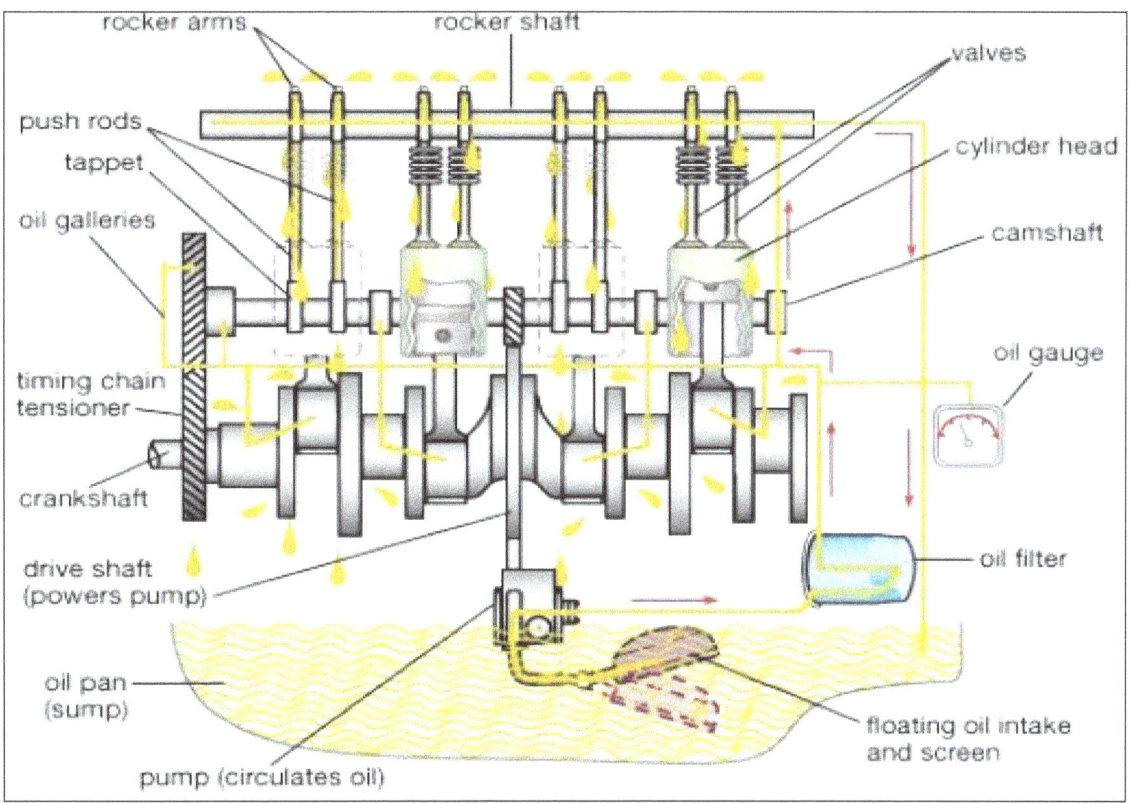

Forced-feed lubrication system.

Nowadays one or a combination of these methods is used to achieve effective cylinder lubrication depending mainly on the operating conditions expected from the engine. Connecting-rod Big-end Side-clearance oil Spray. Cylinder and piston lubrication by big-end side-clearance splash is the simplest and most common method. In this case the oil pressure of the lubrication system, the squeezing action between the connecting-rod and the big-end journal, and the amount of side clearance all together cause sufficient oil splash to the cylinder walls and the underside of the piston when the crankshaft throw is near to the TDC.

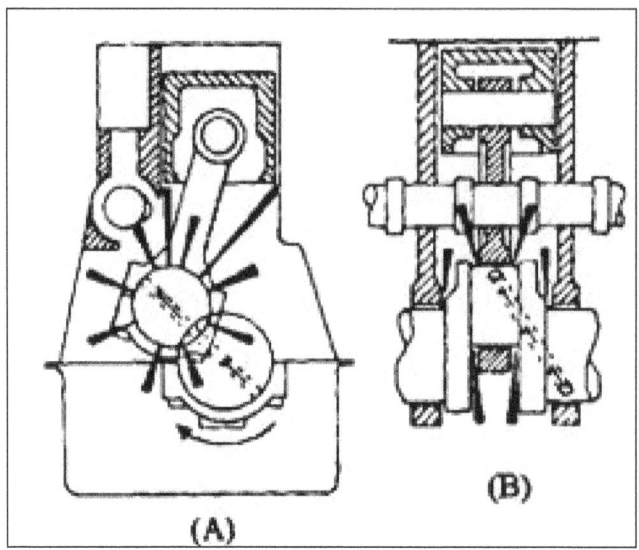

Connecting-rod big-end side-clearance oil spray.

Connecting-rod Big-end Radial-hole Oil Spray

In this case a small radial drilling hole in each connecting-rod big-end directs a squirt of oil to the thrust side of the cylinder bore once in every revolution of the crankshaft. The diameter of the hole and its angular location is critical in this method of lubricating the cylinder.

Connecting-rod Small-end Radial-hole Oil Spray

The connecting-rod in high-performance engines may have a drilling connecting the big-end to the small-end which causes a positive oil feed to the gudgeon-pin. The big-end bearings in heavy-duty diesel engines are grooved so that a continuous flow of oil is provided through the drilled onnect-ing-rods to the small-end bearings. The small-end eyes have two drillings which may supply jets of cooling oil to the ring-belt areas within the pistons.

Crankcase Fixed-jet Oil Spray

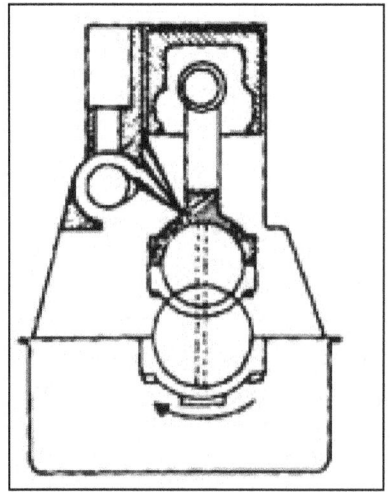

Connecting-rod big-end radial-hole oil spray.

In turbocharged heavy-duty diesel engines, a jet is positioned in the crankcase which projects upwards to provide a controlled and continuous spray of oil that cools and lubricates the underside of the piston. This system of oil supply is more active in reducing piston and ring temperature than providing additional lubrication for the cylinder-and-piston combination.

Small-end Lubrication

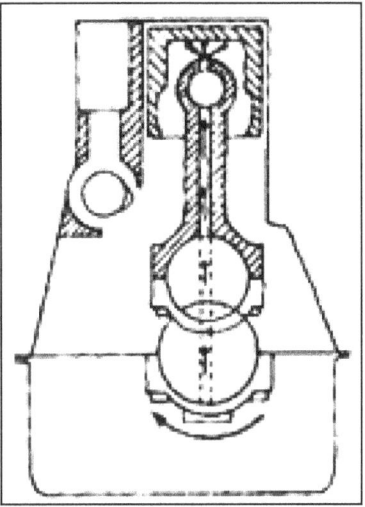

Connecting-rod small-end radial-hole oil spray.

The piston ring scrapes the oil from the cylinder bore on its down-stroke causing pumping action due to which the small-end is positively lubricated. This oil is pushed into the groove behind the lower piston oil-control ring. Subsequently it flows along a drilled passage that intersects the piston gudgeon-pin-boss bores. During pivoting of the small-end of the connecting-rod, this oil spreads over the bearings surfaces. Circumferential slots or drillings are made at right angles to the gudgeon-pin bosses, due to which some of the surplus oil is splashed between the piston gudgeon-pin bosses and the small-end of the connecting-rod. This is necessary when the gudgeon-pin is fully floating and there is no oil supply from the connecting-rod. Additionally, there is a limited amount of splash from the big-end side clearance to complete the small-end lubrication.

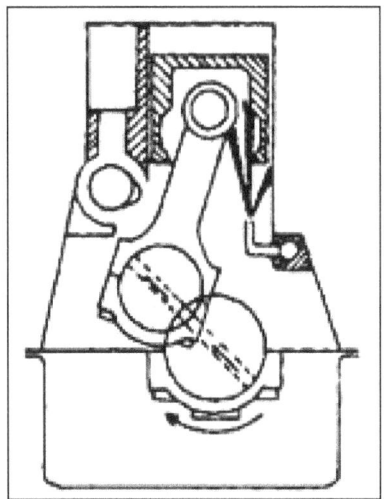

Crankcase fixed-jet oil spray.

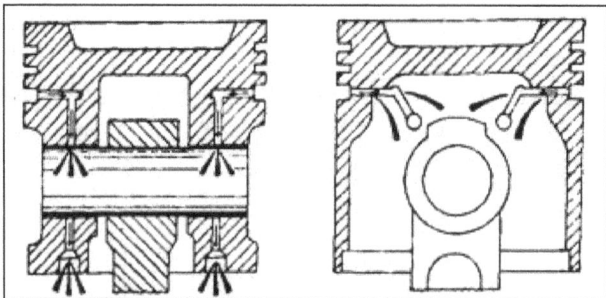

Semi-floating gudgeon-pin with scraper-ring oil supply.

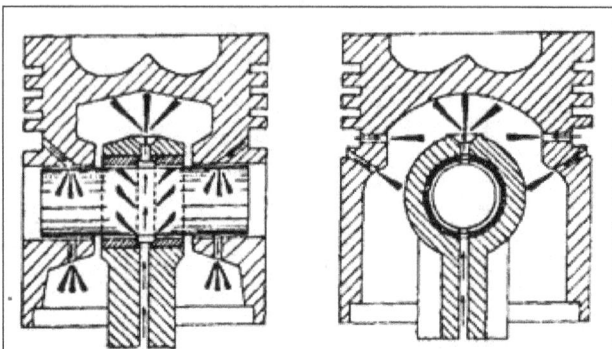

Fully floating gudgeon-pin with scraper-ring and connecting-rod oil supply.

Camshaft-bearing Lubrication

There are four basic methods used for supplying oil to camshaft bearings. (a) Individual cross-drillings in the crankcase directly feed oil from the main oil gallery to each camshaft bearings (b) The drillings in the crankcase, which connect the oil grooves in the crankshaft main bearing to the camshaft bearings, provide a continuous supply of oil. (c) A single drilling provides oil from the main oil gallery to one of the internally grooved camshaft bearings. This oil then enters a pair of radial cross-drillings into the hollow camshaft. A central axial oil passage in the camshaft supplies oil to the other bearings through single radial cross drillings. (d) A separate camshaft oil passage is drilled into and along the length of the camshaft pedestal block. This drilling has intersecting holes connecting it to the various camshaft bearings through which oil is fed.

Camshaft-lobe Profile Lubrication

Methods of lubrication of the camshaft lobes are broadly divided into lubrication for low-mounted camshafts and for high-mounted camshafts. Low-mounted-camshaft lobe lubrication depends mainly on the following: (a) The big-end side clearance allows oil to be flung out, which splashes the cam lobes each time the crank-throw aligns with the cam shaft, i.e., once every revolution of the crankshaft. (b) Draining of oil from the rockers splashes on to the cam profiles. (c) Oil mist is created by the rotating crankshaft and the rocker and crankcase ventilation system.

High-Mounted-Camshaft Lobe Lubrication depends on the type of Valve-Actuating Mechanisms used:

(a) In the direct-acting and centrally pivoted rocker-arm, the cam lobes are provided with a cyclic splash of oil from a drilling in the rocker-arm.

(b) In the direct-acting and end-pivoted rocker-arm, the cam lobes are lubricated by a spray of oil directed on to the lobes. This is provided by a pipe located between the camshaft-bearing pedestal supports.

(c) In direct-acting cylindrical followers, the cam lobes are lubricated by following three methods:

- A simple oil-trough splash.

- A radial drilling intersecting the cam base circle.

- An oil spray coming from a drilled passage along the entire length of the cylinder head.

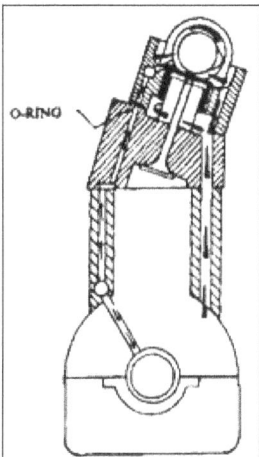

OHCwith directly actuated cylindrical follower.

Poppet-valve Lubrication

The lubrication of the valve stem and tip is carried out by splash of oil and drainage of surplus oil from the rocker-arm and the camshaft lobes.

Valve Rocker-arm-mechanism Lubrication

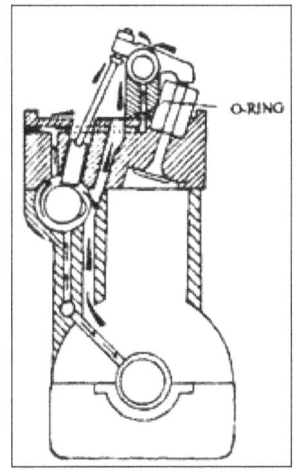 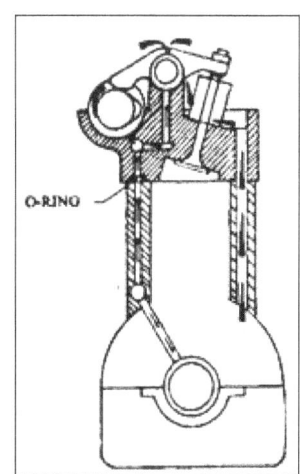

High-mounted camshaft with push rod actuated rocker arm.   OHC with centrally pivoted and end-actuated rocker-arm.

The lubrication of the valve rocker-arm depends on the type of rocker-arm assembly used. Solid Rocker-arm. These arms are lubricated by an oil drilling or pipe extending from one of the camshaft bearings to a hollow rocker-shaft which has radial holes aligning with each rocker-arm. The rocker-arm pivot hole may either be bored and used directly over the shaft or be bronze-bushed with internal oil grooves.

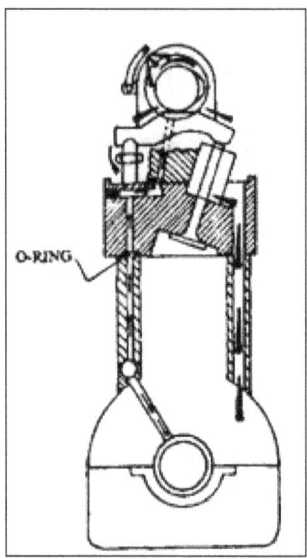

OHC with end-pivoted and centrally actuated rocker-arm.

Three ways of feeding the oil to the valve stem-and-springs assembly and to the tappet and push-rod end are shown in Fig.

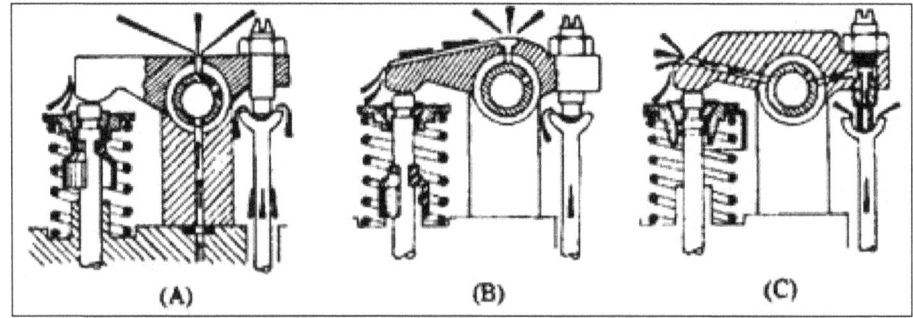

Solid-rocker-arm. oil hole and valve-stem oil-seal arrangements.

(a) The valve stem and the tappet assembly is lubricated by a single vertical radial drilling in the middle of the rocker-arm. As the arm rocks, oil is squirted out in both directions. This method can meet the quantity of oil required for small engines. (b) A more controlled lubrication of the tappet and push-rod assembly is achieved by a horizontal drilling between the rocker-arm pivot hole and the tappet end of the arm. The valve-stem end of the rocker-arm has an open grooved channel formed along the top of the rocker-arm through which the surplus oil floods and drains down over the valve and return-springs. This system is generally adopted on some medium-sized petrol and diesel engines. (c) For heavy-duty operation, lubrication is provided by connecting the rocker-shaft feed to a hollow tappet screw due to which oil flows directly into the push-rod bowl-shaped seat and then overflows and drains down the push-rod lubricating the cam follower. The valve-stem end of the rocker-arm contains a horizontal hole drilled along it so that oil is directly fed to the valve-and-spring assembly. This method, however, may over-lubricate the valve stem if no restriction is imposed on the oil supply to the rockers. This is a problem with this system. One of the camshaft bearings supplies oil through oil drilling to the tappet-follower gallery drilling that lies parallel to the camshaft. From this gallery oil flows around an annular groove in each tappet-follower body ensuring positive lubrication. The flow of this oil through the hollow push-rod and to the rocker-arm and the valve is controlled by a valve disc in the tappet. Oil passes through a passage in the first camshaft bearing to the tappet-follower oil gallery drilled alongside the tappets extending the entire length of the cylinder head. From the gallery oil flows around a recess machined on the tappet and then to a short drilling that meets the central rocker-arm pivot-post stud. The stud is hollow and has a radial intersecting hole so that the oil supply from the tappet gallery is connected to the spherical rocker pivot. The oil then splashes and floods the rocker pressing, consequently overspills lubricating both the valve assembly and the top of the tappet follower.

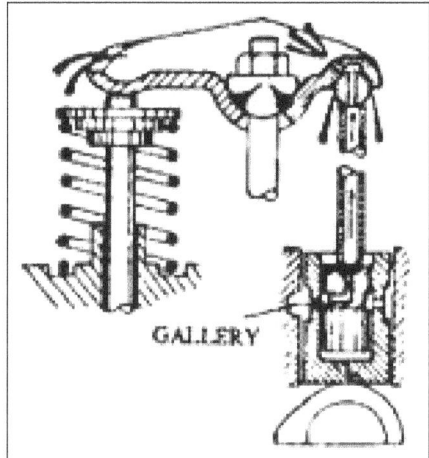

Steel-pressing rocker-arm with hollow-push rod oil supply.

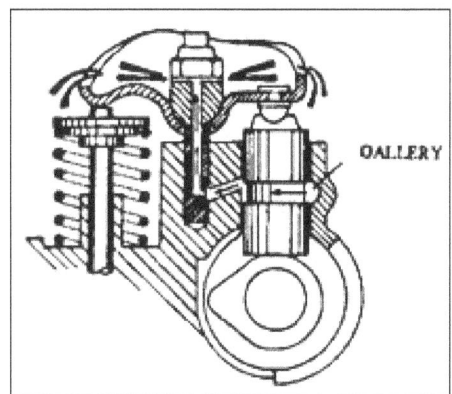

Steel-pressing rocker-arm with central hollow-stud oil supply.

Overhead-camshaft Lubrication

The method of lubrication of overhead camshafts depends on the type of actuating mechanism used and they are as follows: (a) With the direct-acting bucket follower camshaft arrangement, the camshaft, follower, and valve stem are lubricated either by a drilled hole along the centre of the camshaft axis and intersecting radial holes emerging on the base circle of the cam, or by a drilled hole in the pedestal casing parallel to the camshaft and projecting spray holes directed on to the cam profiles. In either of the cases the follower and the valve stem are lubricated by drainage of oil from the camshaft. (b) With the indirect end-pivoted rocker-arm arrangement, the camshaft, rocker, and valve stem are lubricated by spraying oil on to the cam faces through an oil pipe attached to the camshaft pedestal housing. The excess oil draining from the camshaft also flows over the rocker-arm and lubricates its pivot joint and the valve tip and stem.

Lubrication of Timing Gears and Chains

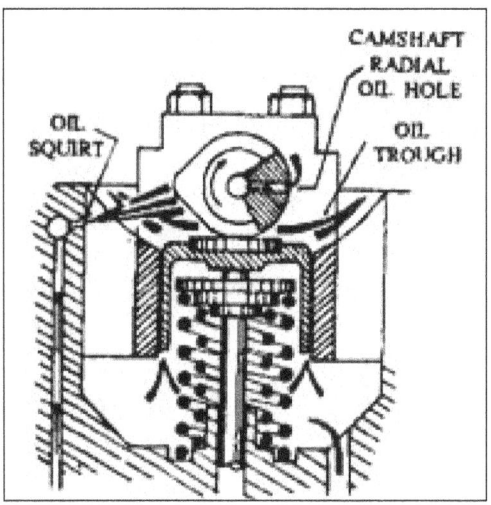

OHC cylindrical direct-acting with a fixed-pedestal spray, a hollow camshaft
with a radial oil hole, or simply a trough splash bath.

These excessively used components are normally lubricated by a small drilling, which intersects the oil passage running from the main oil gallery to the first main bearing or the passage from the first main bearing to the first camshaft bearing. Sometimes a small pipe from this drilling directs the oil on to the gears or chain. Moreover, in some constructions the sump is shaped to form a timing-gear oil trough, due to which the draining oil submerges the crankshaft gear providing a continuous upward oil splash to the rest of the camshaft drive.

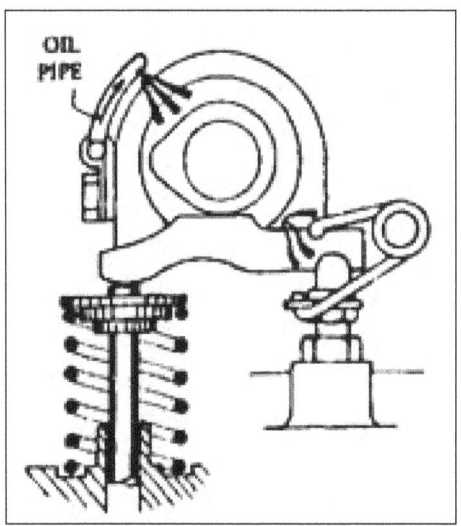

OHC with end-pivoted rocker-arm and oil-pipe-supply spray.

Crankshaft Oil Passages

Crankshaft oil passages feed oil from the main-journal bearing to the big-end journal. In its simplest form, the oil passage is a diagonal drilling running from the main journal to the big-end journal. Normally the diagonal hole is drilled at an angle to the crank-web centre-line so that, when the crank-pin is in the TDC position and combustion force pushes the connecting rods downwards, some oil still enters between the journal and the bearing. It is because if the exit of the diagonal hole is exactly at the top of the big-end journal, oil can not enter between the bearing and the journal in the TDC position. Additionally the effective projected bearing area is also reduced by chamfered oil hole. To have an improvement in oil delivery, a cross-drilling runs straight through the big-end journal and a diagonal drilling from the main-bearing journal intersects the big-end cross-drilling. Another hole is also drilled diametrically opposite the diagonal-hole's entry in the main journal, so that when the bearing is loaded at the top or the bottom of the stroke, the other side of the bearing permits oil to enter.

Oil Pumps

Four basic types of rotary-operating oil-pumps are used in pressure-feed lubrication systems. They are:

(a) External-spur-gear pump, (b) Internal-gear crescent pump, (c) Eccentric bi-rotor pump, and (d) Sliding-vane eccentric pump.

Selection of a pump is usually based on their ease and convenience of being installed and driven. Other considerations are expected pump life, oil-flow rate capacity, priming time, the ability to built up pressure at low speeds, and the ability to deliver oil under higher pressure conditions continuously at high engine speeds.

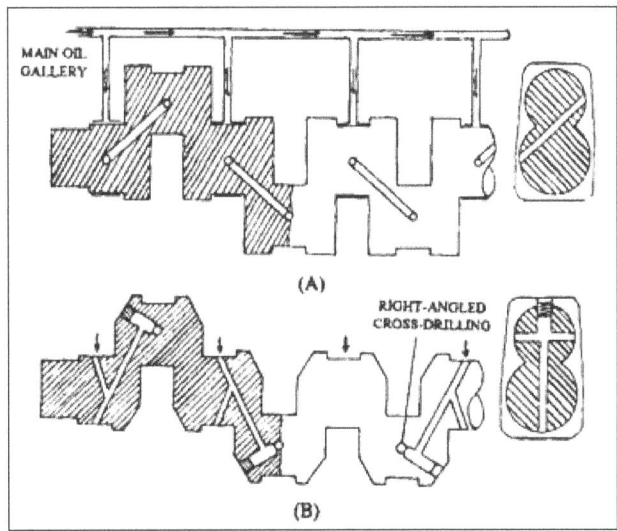

Crankshaft oil passages.

A. Crankshaft with single oil passage.

B. Crankshaft with diagonal web passage and right-angled cross-drilling in the big-end journal.

The generation of oil pressure by the oil pump depends on the "leaks" in the engine. The "leaks" are the clearances at end points of the lubrication system, such as the edges of bearings, the rocker arms, the connecting rod spit holes, etc. These clearances are introduced for proper operations of the engine. The leakage increases as parts wear and clearance becomes greater. The oil pumps capacity is based on its size, rotating speed, and physical condition. The oil pump capacity is low at engine idling and when the "leaks" are relatively more than the pump capacity. As the engine speeds up, the pump capacity increases and it tries to force more oil out of the "leaks". This causes the pressure to rise until it reaches the regulated pressure. Engine oil viscosity also takes part in both the pump capacity and the oil leakage. Very low viscosity or thin oil slips past the edges of the pump and flows freely from the "leaks". Hot oil has low viscosity and, is often accompanied by low oil pressure. Cold oil is more viscous and usually causes high pressures, even with the engine idling. Higher viscosity oil in an engine raises the oil pressure even to the regulated value at a lower engine speed.

External-spur-gear Pump

This pump is consisted of two identical meshing spur gearwheels installed in the pump body. The driving gearwheel is rigidly connected to the oil pump drive shaft by shrunk fit or by a keyway. The drive shaft rotates in a bearing bore machined directly in the pump housing, and the driven gear revolves on a bearing post mounted within this housing. As the gears rotate a low pressure area is produced on the inlet suction side so that oil is drawn in. The oil filling the spaces between the gear teeth is sealed off by the housing walls as the wheels rotate. This trapped oil then moves around the periphery of each gearwheel, in opposite directions in the two gearwheels, to the discharge outlet port. The continuous displacement of oil to the outlet port pressurises the oil and increases the rate of oil circulation. A single stage pump of this type can develop the delivery pressure up to 981 kPa. The discharging capacity of the pump can be determined using the following relation:

$$Q = \frac{2alnN}{60 \times 1000} \, l/hr$$

where Q = pump discharge, l/hr

   a = area enclosed between two adjacent teeth, cm^2

   l = axial length of teeth, cm

   n = number of teeth in each gear

   N = speed of the gears, rpm.

The pump components are checked for correct clearance and wear using feeler gauges. The clearance between the gear tooth tips and the pump body should not exceed 0.2 mm. The backlash between the meshing gears should be between 0.1 and 0.2 mm. The end float clearance (between the end-plate and the gears across the open face of the pump casing) from the gear faces should not exceed 0.1 mm. Mechanical and volumetric efficiencies of these pumps are quite high being 95% and 98% respectively.

Internal-gear Crescent Pump

This pump contains an internal-spur ring gear that runs outside but in mesh with a driving external-spur gear, in such a way that its axis of rotation is eccentric to that of the driving gear. This eccentricity causes a space between the external and internal gears. This space is occupied by a fixed spacer block called the crescent the purpose of which is to separate the inlet and output port areas. The driving gear is driven either by a separate shaft or is keyed to an extension of the front crankshaft main journal. The outer-gear axis of rotation is maintained entirely by the pump casing wall. The housing of the pump around a crankshaft journal permits a very compact pump unit, capable of delivering large flow output at relatively low crankshaft speeds.

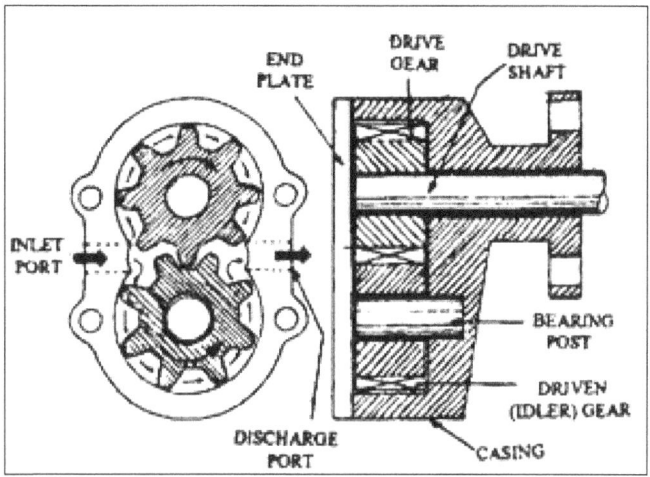

Spur external-gear pump.

The rotation of the gears develops a low-pressure area at the inlet suction end of the crescent so that oil is drawn in. As the gearwheels rotate, oil is trapped between teeth of the inner driver gear and the inside crescent side wall, and between teeth of the outer gear and the outside crescent side wall. This oil is carried round by these teeth to the other end of the crescent, where it is discharged by both sets of teeth into the outlet port chamber. The continuous supply of oil to the outlet side increases oil discharge. Once the space between the gear teeth has been filled with oil, the extra oil squeezed out from the teeth gaps increases the oil pressure. The mean oil pressure and the rate of circulation depend on the amount of oil escaping from the lubrication system's bearings. The pump components are checked for correct clearance and wear using feeler gauges. The clearances between the gear tooth tips and the crescent wall for each gear should not exceed 0.3 mm when both gears are is position with the pump body. The backlash between the meshing gears should be between 0.1 and 0.2 mm. The clearance between the outer gear and the body should not exceed 0.2 mm. The clearance between gear end-float across the open face of the pump casing and the gear side faces should not exceed 0.2 mm.

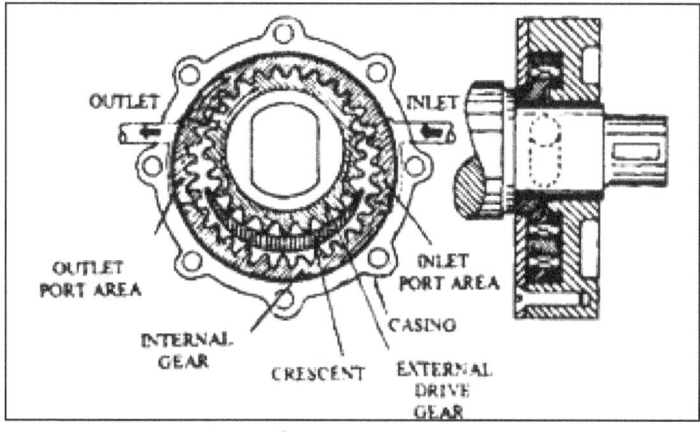

Internal-gear crescent pump.

Eccentric Bi-rotor Pump

The pump uses an inner and an outer rotor installed in the pump body, and the outer rotor is eccentric to the inner. The inner rotor is pressed on to the oil-pump shaft and is held in position by serrations. This rotor has four lobes which mesh with five segments in the outer rotor. The inner rotor thus revolves the outer rotor, but at a speed which is slower by the ratio of the number of lobes to segments. First oil is drawn, through the inlet port, into the space between the inner and outer rotors. Due to their eccentricity and difference in size, the gap between the lobes increases and consequently oil is filled up in this space. However the space between the rotor lobes moves beyond the inlet port, thus trapping the oil, which is subsequently carried around between the rotor lobes and segments. With further rotation, the volume of this effective space formed decreases and it is eventually exposed to the delivery port, so that the oil is discharged under pressure to the filter. The pump acts by continuous repetition of this process.

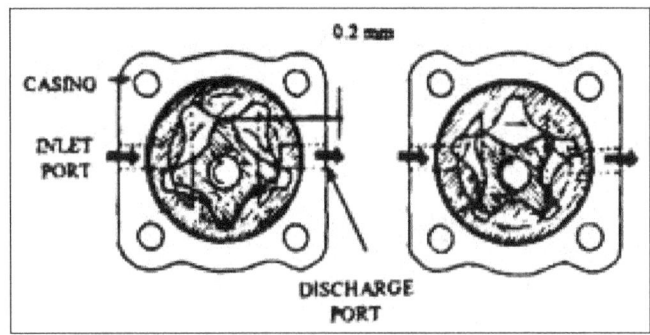
Eccentric bi-rotor pump.

The pump components are checked for correct clearance and wear using feeler gauges. The clearance between the rotor lobe tip and the segment should not exceed 0.2 mm. The clearance between the outer rotor and the body should not exceed 0.25 mm. The end-float clearance (between the end-plate and the rotors) across the open face of the pump casing and the rotors for the internal-gear crescent pump should not exceed 0.2 mm.

Sliding-vane Eccentric Pump

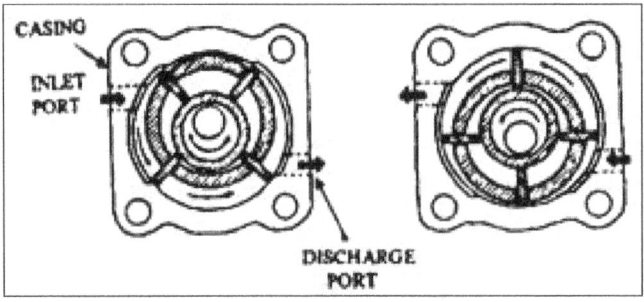
Sliding-vane pump.

This pump contains a rotor installed eccentrically in a cylinder bore machined in the pump body. The rotor is pressed on to the oil-pump shaft and is positively retained by a pin. Four sliding vanes are placed in grooves machined in the periphery of this rotor and are located by centralising rings on each side of the rotor. During the operation of the pump, the vanes are held against the pump-body wall by centrifugal force. As the rotor moves, the vanes pass over the inlet port formed in the side of the pump body. Due to the eccentricity of the rotor shaft to the casing wall, the space between the vanes in-creases and oil is drawn into the space between the rotor and the pump-body wall. The oil is subsequently carried round between the vanes beyond the inlet port, where the space between the rotor and the pump bore decreases. Consequently the oil is forced out through the discharge port to the oil filter and oil galleries. Due to the displacement of the excessive quantity of oil further, the oil pressure in the engine's lubrication passages increases.

The pump components are checked for correct clearance and wear using feeler gauges. The clearance between the rotor and the body should not exceed 0.13 mm. The clearance between the vane and the body should not exceed 0.28 mm. The clearance between the vane and the rotor groove should not exceed 0.13 mm. The rotor and vane end-float clearance across the open face of the pump body from the rotor and vanes should not exceed 0.13 mm.

Oil-pump Drive Arrangements

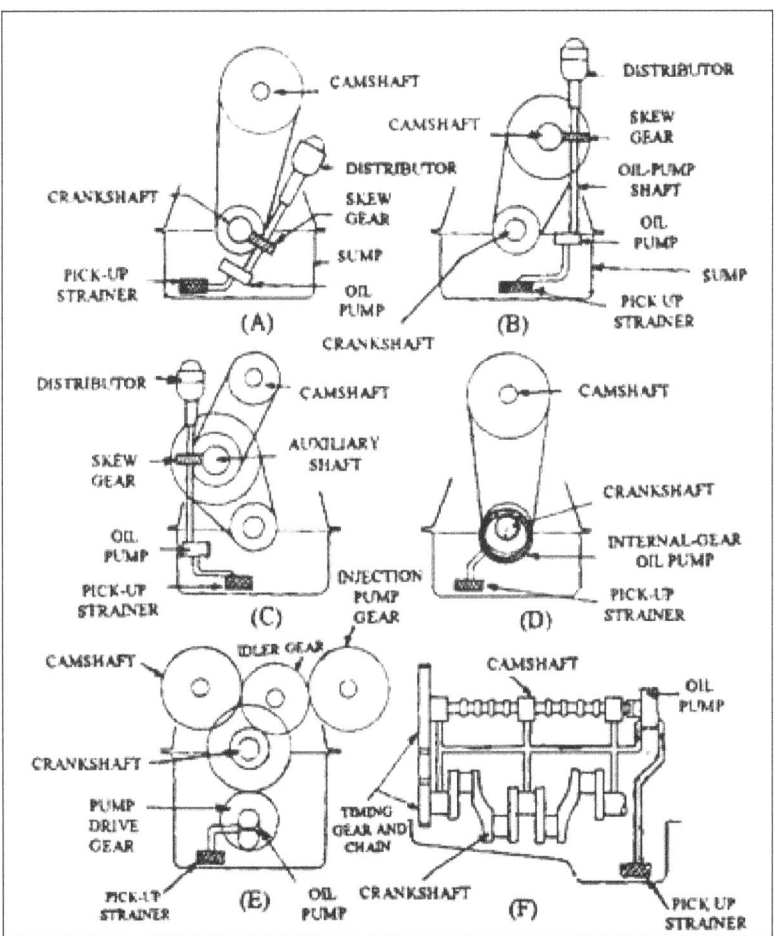

Oil-pump drive arrangements. A. Crankshaft skew-gear drive. B. Camshaft skew-gear drive. C. Auxiliary-shaft skew drive. D. Crankshaft internal-gear drive. E. Crankshaft timing-gear drive. F. Camshaft direct end couple drives.

Skew gear drive for crankshaft oil pump is employed when high mounted camshafts are used. A short shaft with a skew gear meshes with a gear mounted on the front of the crankshaft. This arrangement drives both the ignition distributor and the oil pump with a 2:1 gear reduction Skew gear drive for camshaft oil-pump is used when a low mounted camshaft is employed. A long shaft with a skew gear meshes with a gear machined directly on the camshaft. This shaft drives both the ignition distributor and the oil-pump with a 1: 1 gear ratio. When a double-stage timing chain is employed with a high-mounted camshaft and an auxiliary shaft (also known as a jack shaft), it is often convenient to have the oil pump drive from this shaft. This also drives both the distributor and the pump. For small compact engines with high mounted camshafts, an internal-gear crescent pump is commonly used. This pump is compact, occupies very little space and sits over a keyed external gear on the crankshaft. This serves dual function of driving the pump and generating part of the pumping action. For medium-and-large sized commercial engines, driving the pump directly from the crankshaft timing gear is preferred. These forms of drive are commonly used with large output pumps. For some transverse-mounted engines with low-mounted camshafts, it is preferred to drive a pump through a coupling located at the end of the camshaft. This is also compact and dispenses with a separate drive shaft.

Pressure Regulator

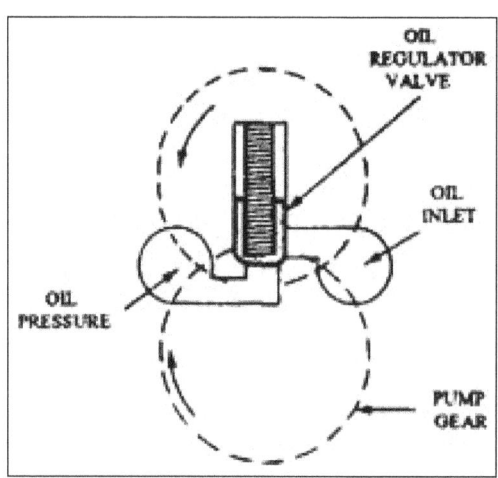

Oil pressure regulator valve.

A pressure regulator or relief valve limits the maximum pressure in engines with full pressure lubricating systems. If a pressure regulator valve is not used, the engine oil pressure continues to increase as the engine speed increases. Maximum pressure is usually limited to the pressure that delivers an adequate quantity of oil (11 to 22 1pm) to engine lubricating points. After oil leaves the pump, oil films are maintained by hydrodynamic forces. Excessive oil pressure requires more power and also does not provide better lubrication. High oil pressure and consequently the resulting high rates of oil flow may cause erosion to engine bearings in some cases. The pressure regulator is installed downstream from the other pressure side of the oil pump. It normally contains a spring-loaded piston and in some cases, a spring-loaded ball. When oil pressure reaches regulated pressure (normally between 380 to 414 kPa), it forces the regulator valve back against the calibrated spring, compressing it as the valve is forced back. This allows a controlled "leak" from the pressure system so that the regulated set pressure is maintained. Any change in the regulator valve spring pressure also changes the regulated oil pressure and higher spring pressure causes higher oil pressure. In most engines, released oil from regulator valve is routed to the inlet side of the pump as shown in Fig. The regulator valve is, therefore, usually placed in the oil pump housing or pump cover. This method of oil flow from regulator valve prevents foaming and excessive oil agitation so that the pump receives a solid stream of lubricating oil.

Oil Filter

Oil from the pump outlet flows to the oil filter where large particles are trapped, allowing only clean oil to flow into the engine. Very fine particles flow through the filter. These particles are so fine they can get between engine clearances causing no damage. As the filter traps particles, the holes in the filter become partly plugged so that it traps even smaller particles. This better filtering, however, restricts oil flow which can result in bearing oil starvation. Most oil filters are the spin-on, disposable type. Oil flows from the oil pump into the outside area of the filter between the filter case and the paper element. The oil flows from the centre of the paper element to the main oil gallery in the engine block. A check valve placed in the top of the oil filter prevents oil from draining back out of the lubrication system through the filter into the crankcase when the engine is shut off. The oil filter mounting plate contains a bypass valve which allows unfiltered oil to flow from the oil pump directly to the lubrication system if the filter becomes plugged. This bypass valve is set at from 34 to 103 kPa, depending on the engine and normal pressure drop across the filter element.

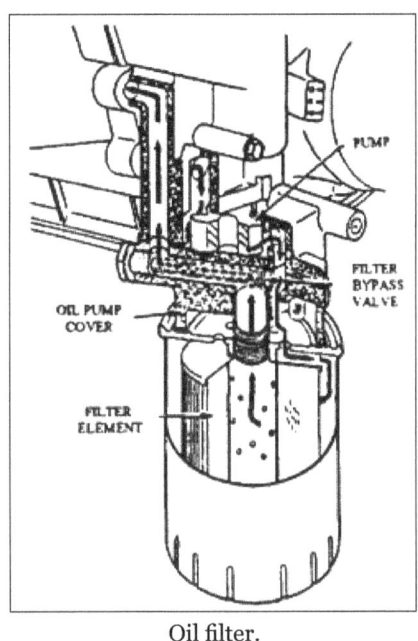

Oil filter.

Lubrication System for Diesel Engines

The lubrication system in a diesel engine is very similar to the systems in gasoline engines. However, there are some significant differences between the two systems. An oil cooler is connected in series with the oil filter on some diesel engines. Engine coolant is circulated around the oil passages in the cooler to dissipate heat from the oil. Oil is supplied from passages in the engine block to curved discharge tubes at the bottom of each cylinder. The oil is sprayed from an orifice in the discharge tubes against the under crown of the piston to provide additional piston cooling and more uniform piston temperature. This extra piston cooling allows close clearances between the piston rings and ring grooves and reduces the possibility of the rings sticking in their grooves. Figure shows the oil-cooled piston features.

Also high combustion temperature and elements such as sulphur in the diesel fuel make the CI engines more prone to deposits of carbon, formation of gums around the piston rings, and lacquer deposits on the side of the pistons. These conditions are minimised by using heavy-duty detergent oil. Detergent additives hold the carbon and soot in suspension in the oil so that the foreign particles harmlessly pass around the system until the oil is next drained.

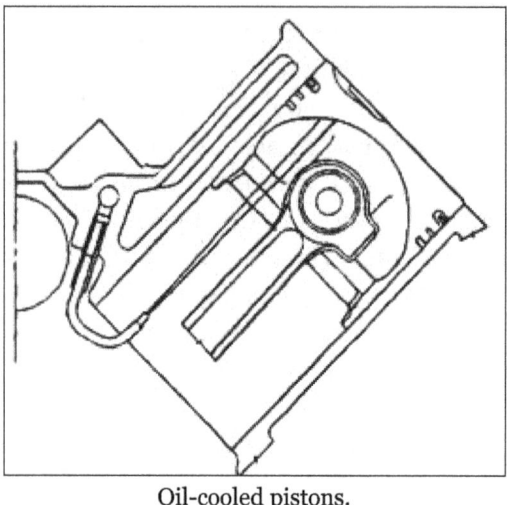

Oil-cooled pistons.

AUTOMOTIVE ELECTRONICS

Automotive electronics are electronic systems used in vehicles, including engine management, ignition, radio, carputers, telematics, in-car entertainment systems, and others. Ignition, engine and transmission electronics are also found in trucks, motorcycles, off-road vehicles, and other internal combustion powered machinery such as forklifts, tractors and excavators. Related elements for control of relevant electrical systems are also found on hybrid vehicles and electric cars.

Electronic systems have become an increasingly large component of the cost of an automobile, from only around 1% of its value in 1950 to around 30% in 2010. Modern electric cars rely on power electronics for the main propulsion motor control, as well as managing the battery system. Future autonomous cars will rely on powerful computer systems, an array of sensors, networking, and satellite navigation, all of which will require electronics.

Types

Automotive electronics or automotive embedded systems are distributed systems, and according to different domains in the automotive field, they can be classified into:

- Engine electronics.

- Transmission electronics.

- Chassis electronics.

- Passive safety.

- Driver assistance.

- Passenger comfort.

- Entertainment systems.

- Electronic integrated cockpit systems.

Engine Electronics

One of the most demanding electronic parts of an automobile is the engine control unit (ECU). Engine controls demand one of the highest real-time deadlines, as the engine itself is a very fast and complex part of the automobile. Of all the electronics in any car, the computing power of the engine control unit is the highest, typically a 32-bit processor.

A modern car may have up to 100 ECU's and a commercial vehicle up to 40.

An engine ECU controls such functions as,

In a diesel engine:

- Fuel injection rate.

- Emission control, NOx control.

- Regeneration of oxidation catalytic converter.

- Turbocharger control.

- Cooling system control.

- Throttle control.

In a gasoline engine:

- Lambda control.

- OBD (On-Board Diagnostics).

- Cooling system control.

- Ignition system control.

- Lubrication system control (only a few have electronic control).

- Fuel injection rate control.

- Throttle control.

Many more engine parameters are actively monitored and controlled in real-time. There are about 20 to 50 that measure pressure, temperature, flow, engine speed, oxygen level and NOx level plus other parameters at different points within the engine. All these sensor signals are sent to the ECU, which has the logic circuits to do the actual controlling. The ECU output is connected to different actuators for the throttle valve, EGR valve, rack (in VGTs), fuel injector (using a pulse-width modulated signal), dosing injector and more. There are about 20 to 30 actuators in all.

Transmission Electronics

These control the transmission system, mainly the shifting of the gears for better shift comfort and to lower torque interrupt while shifting. Automatic transmissions use controls for their operation, and also many semi-automatic transmissions having a fully automatic clutch or a semi-auto clutch (declutching only). The engine control unit and the transmission control exchange messages, sensor signals and control signals for their operation.

Chassis Electronics

The chassis system has a lot of sub-systems which monitor various parameters and are actively controlled:

- ABS - Anti-lock Braking System.

- TCS – Traction Control System.

- EBD – Electronic Brake Distribution.

- ESP – Electronic Stability Program.

- PA - Parking Assistance.

Passive Safety

These systems are always ready to act when there is a collision in progress or to prevent it when it senses a dangerous situation:

- Air bags.

- Hill descent control.

- Emergency brake assist system.

Driver Assistance

- Lane assist system.

- Speed assist system.

- Blind spot detection.

- Park assist system.

- Adaptive cruise control system.

- Pre-collision Assist.

Passenger Comfort

- Automatic climate control.

- Electronic seat adjustment with memory.

- Automatic wipers.

- Automatic headlamps - adjusts beam automatically.

- Automatic cooling - temperature adjustment.

Entertainment Systems

- Navigation system.

- Vehicle audio.

- Information access.

All of the above systems form an infotainment system. Developmental methods for these systems vary according to each manufacturer. Different tools are used for both hardware and software development.

Electronic Integrated Cockpit Systems

These are new generation hybrid ECUs that combine the functionalities of multiple ECUs of Infotainment Head Unit, Advanced Driver Assistance Systems (ADAS), Instrument Cluster, Rear Camera/Parking Assist, Surround View Systems etc. This saves on the cost of electronics as well as mechanical/physical parts like interconnects across ECUs etc. There is also a more centralized control so data can be seamlessly exchanged between the systems.

There are of course challenges too. Given the complexity of this hybrid system, a lot more rigor is needed to validate the system for robustness, safety and security. For example, if the infotainment system's application which could be running an open-source Android OS is breached, there could be possibility of hackers to take control of the car remotely and potentially misuse it for anti-social activities. Typically so, usage of a hardware+software enabled hypervisors are used to virtualize and create separate trust and safety zones that are immune to each other's failures or breaches.

Functional Safety Requirements

In order to minimize the risk of dangerous failures, safety-related electronic systems have to be developed following the applicable product liability requirements. Disregard for, or inadequate application of these standards can lead to not only personal injuries, but also severe legal and economic consequences such as product cancellations or recalls.

The IEC 61508 standard, generally applicable to electrical/electronic/programmable safety-related products, is only partially adequate for automotive-development requirements. Consequently, for the automotive industry, this standard is replaced by the existing ISO 26262, currently released as a Final Draft International Standard (FDIS). ISO/DIS 26262 describes the entire product life-cycle of safety-related electrical/electronic systems for road vehicles. It has been published as an international standard in its final version in November 2011. The implementation of this new standard will result in modifications and various innovations in the automobile electronics development process, as it covers the complete product life-cycle from the concept phase until its decommissioning.

Security

As more functions of the automobile are connected to short- or long-range networks, cybersecurity of systems against unauthorized modification is required. With critical systems such as engine controls, transmission, airbags, and braking connected to internal diagnostic networks, remote access could result in a malicious intruder altering the function of systems or disabling them, possibly causing injuries or fatalities. Every new interface presents a new "attack surface". The same facility that allows the owner to unlock and start a car from a smartphone app also presents risks due to remote access. Auto manufacturers may protect the memory of various control microprocessors both to secure them from unauthorized changes and also to ensure only manufacturer-authorized facilities can diagnose or repair the vehicle. Systems such as keyless entry rely on cryptographic techniques to ensure "replay" or "man-in-the-middle attacks" attacks cannot record sequences to allow later break-in to the automobile.

COOLING SYSTEM

A typical 4 cylinder vehicle cruising along the highway at around 50 miles per hour, will produce 4000 controlled explosions per minute inside the engine as the spark plugs ignite the fuel in each cylinder to propel the vehicle down the road. Obviously, these explosions produce an enormous amount of heat and, if not controlled, will destroy an engine in a matter of minutes. Controlling these high temperatures is the job of the cooling system.

The modern cooling system has not changed much from the cooling systems in the model T back in the '20s. Oh sure, it has become infinitely more reliable and efficient at doing it's job, but the basic cooling system still consists of liquid coolant being circulated through the engine, then out to the radiator to be cooled by the air stream coming through the front grill of the vehicle.

Today's cooling system must maintain the engine at a constant temperature whether the outside air temperature is 110 degrees Fahrenheit or 10 below zero. If the engine temperature is too low, fuel economy will suffer and emissions will rise. If the temperature is allowed to get too hot for too long, the engine will self destruct.

Workings of a Cooling System

Actually, there are two types of cooling systems found on motor vehicles: Liquid cooled and Air cooled. Air cooled engines are found on a few older cars, like the original Volkswagen Beetle, the Chevrolet Corvair and a few others.

The cooling system is made up of the passages inside the engine block and heads, a water pump to circulate the coolant, a thermostat to control the temperature of the coolant, a radiator to cool the coolant, a radiator cap to control the pressure in the system, and some plumbing consisting of interconnecting hoses to transfer the coolant from the engine to radiator and also to the car's heater system where hot coolant is used to warm up the vehicle's interior on a cold day.

A cooling system works by sending a liquid coolant through passages in the engine block and heads. As the coolant flows through these passages, it picks up heat from the engine. The heated fluid then makes its way through a rubber hose to the radiator in the front of the car. As it flows through the thin tubes in the radiator, the hot liquid is cooled by the air stream entering the engine compartment from the grill in front of the car. Once the fluid is cooled, it returns to the engine to absorb more heat. The water pump has the job of keeping the fluid moving through this system of plumbing and hidden passages.

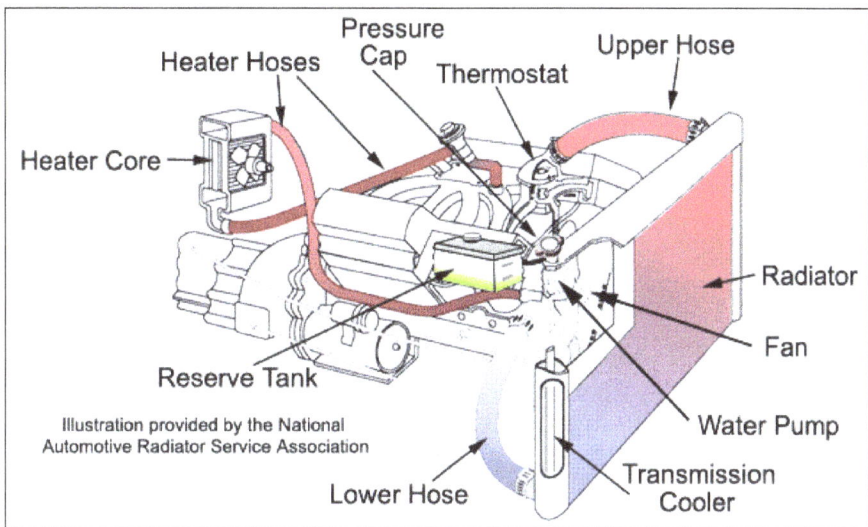

A thermostat is placed between the engine and the radiator to make sure that the coolant stays above a certain preset temperature. If the coolant temperature falls below this temperature, the thermostat blocks the coolant flow to the radiator, forcing the fluid instead through a bypass directly back to the engine. The coolant will continue to circulate like this until it reaches the design temperature, at which point, the thermostat will open a valve and allow the coolant back through the radiator.

In order to prevent the coolant from boiling, the cooling system is designed to be pressurized. Under pressure, the boiling point of the coolant is raised considerably. However, too much pressure will cause hoses and other parts to burst, so a system is needed to relieve pressure if it exceeds a certain point. The job of maintaining the pressure in the cooling system belongs to the radiator cap. The cap is designed to release pressure if it reaches the specified upper limit that the system was designed to handle. Prior to the '70s, the cap would release this extra pressure to the pavement. Since then, a system was added to capture any released fluid and store it temporarily in a reserve tank. This fluid would then return to the cooling system after the engine cooled down. This is what is called a closed cooling system.

Circulation

The coolant follows a path that takes it from the water pump, through passages inside the engine block where it collects the heat produced by the cylinders. It then flows up to the cylinder head (or heads in a V type engine) where it collects more heat from the combustion chambers. It then flows out past the thermostat (if the thermostat is opened to allow the fluid to pass), through the upper radiator hose and into the radiator. The coolant flows through the thin flattened tubes that make up the core of the radiator and is cooled by the air flow through the radiator. From there, it flows out of the radiator, through the lower radiator hose and back to the water pump. By this time, the coolant is cooled off and ready to collect more heat from the engine.

The capacity of the system is engineered for the type and size of the engine and the work load that it is expected to undergo. Obviously, the cooling system for a larger, more powerful V8 engine in a heavy vehicle will need considerably more capacity then a compact car with a small 4 cylinder engine. On a large vehicle, the radiator is larger with many more tubes for the coolant to flow through. The radiator is also wider and taller to capture more air flow entering the vehicle from the grill in front.

Antifreeze

The coolant that courses through the engine and associated plumbing must be able to withstand temperatures well below zero without freezing. It must also be able to handle engine temperatures in excess of 250 degrees without boiling. A tall order for any fluid, but that is not all. The fluid must also contain rust inhibiters and a lubricant.

The coolant in today's vehicles is a mixture of ethylene glycol (antifreeze) and water. The recommended ratio is fifty-fifty. In other words, one part antifreeze and one part water. This is the minimum recommended for use in automobile engines. Less antifreeze and the boiling point would be too low. In certain climates where the temperatures can go well below zero, it is permissible to have as much as 75% antifreeze and 25% water, but no more than that. Pure antifreeze will not work properly and can cause a boil over.

Antifreeze is poisonous and should be kept away from people and animals, especially dogs and cats, who are attracted by the sweet taste. Ethylene Glycol, if ingested, will form calcium oxalate crystals in the kidneys which can cause acute renal failure and death.

The Components of a Cooling System

- The Radiator.

- Radiator Cooling Fans.

- Pressure Cap & Reserve Tank.

- Water Pump.

- Thermostat.

- Bypass System.

- Freeze Plugs.

- Head Gaskets & Intake Manifold Gaskets.

- Heater Core.

- Hoses.

The Radiator

The radiator core is usually made of flattened aluminum tubes with aluminum strips that zigzag between the tubes. These fins transfer the heat in the tubes into the air stream to be carried away from the vehicle. On each end of the radiator core is a tank, usually made of plastic that covers the ends of the radiator.

On most modern radiators, the tubes run horizontally with the plastic tank on either side. On other cars, the tubes run vertically with the tank on the top and bottom. On older vehicles, the core was made of copper and the tanks were brass. The new aluminum-plastic system is much more efficient, not to mention cheaper to produce. On radiators with plastic end caps, there are gaskets between the aluminum core and the plastic tanks to seal the system and keep the fluid from leaking out. On older copper and brass radiators, the tanks were brazed (a form of welding) in order to seal the radiator.

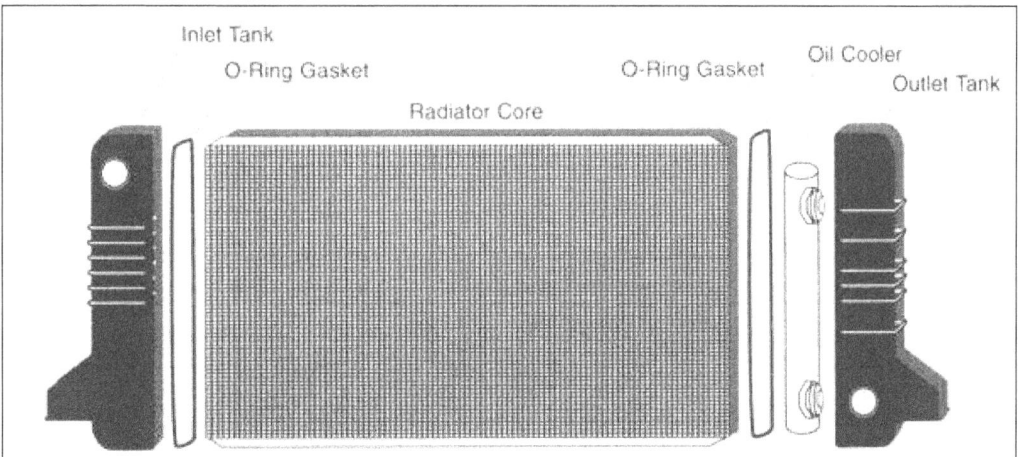

The tanks, whether plastic or brass, each have a large hose connection, one mounted towards the top of the radiator to let the coolant in, the other mounted at the bottom of the radiator on the other tank to let the coolant back out. On the top of the radiator is an additional opening that is capped off by the radiator cap.

Another component in the radiator for vehicles with an automatic transmission is a separate tank mounted inside one of the tanks. Fittings connect this inner tank through steel tubes to the automatic transmission. Transmission fluid is piped through this tank inside a tank to be cooled by the coolant flowing past it before returning the the transmission.

Radiator Fans

Mounted on the back of the radiator on the side closest to the engine is one or two electric fans inside a housing that is designed to protect fingers and to direct the air flow. These fans are there to keep the air flow going through the radiator while the vehicle is going slow or is stopped with the engine running. If these fans stopped working, every time you came to a stop, the engine temperature would begin rising. On older systems, the fan was connected to the front of the water pump and would spin whenever the engine was running because it was driven by a fan belt instead of an electric motor. In these cases, if a driver would notice the engine begin to run hot in stop and go driving, the driver might put the car in neutral and rev the engine to turn the fan faster which helped cool the engine. Racing the engine on a car with a malfunctioning electric fan would only make things worse because you are producing more heat in the radiator with no fan to cool it off.

The electric fans are controlled by the vehicle's computer. A temperature sensor monitors engine temperature and sends this information to the computer. The computer determines if the fan should be turned on and actuates the fan relay if additional air flow through the radiator is necessary.

If the car has air conditioning, there is an additional radiator mounted in front of the normal radiator. This "radiator" is called the air conditioner condenser, which also needs to be cooled by the air flow entering the engine compartment. As long as the air conditioning is turned on, the system will keep the fan running, even if the engine is not running hot. This is because if there is no air flow through the air conditioning condenser, the air conditioner will not be able to cool the air entering the interior.

Pressure Cap and Reserve Tank

As coolant gets hot, it expands. Since the cooling system is sealed, this expansion causes an increase in pressure in the cooling system, which is normal and part of the design. When coolant is under pressure, the temperature where the liquid begins to boil is considerably higher. This pressure, coupled with the higher boiling point of ethylene glycol, allows the coolant to safely reach temperatures in excess of 250 degrees.

The radiator pressure cap is a simple device that will maintain pressure in the cooling system up to a certain point. If the pressure builds up higher than the set pressure point, there is a spring loaded valve, calibrated to the correct Pounds per Square Inch (psi), to release the pressure.

When the cooling system pressure reaches the point where the cap needs to release this excess pressure, a small amount of coolant is bled off. It could happen during stop and go traffic on an extremely hot day, or if the cooling system is malfunctioning. If it does release pressure under these conditions, there is a system in place to capture the released coolant and store it in a plastic tank that is usually not pressurized. Since there is now less coolant in the system, as the engine cools down a partial vacuum is formed. The radiator cap on these closed systems has a secondary valve to allow the vacuum in the cooling system to draw the coolant back into the radiator from the reserve tank (like pulling the plunger back on a hypodermic needle) There are usually markings on the side of the plastic tank marked Full-Cold, and Full Hot. When the engine is at normal operating temperature, the coolant in the translucent reserve tank should be up to the Full-Hot line. After the engine has been sitting for several hours and is cold to the touch, the coolant should be at the Full-Cold line.

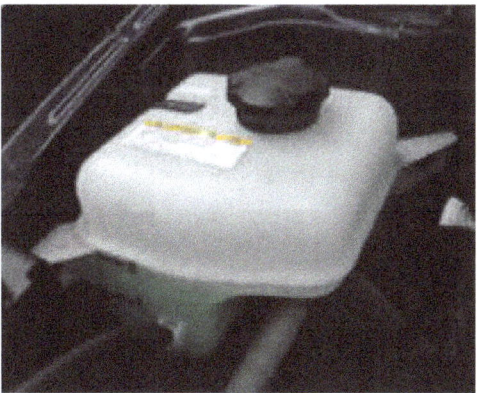

Water Pump

A water pump is a simple device that will keep the coolant moving as long as the engine is running. It is usually mounted on the front of the engine and turns whenever the engine is running. The water pump is driven by the engine through one of the following:

- A fan belt that will also be responsible for driving an additional component like an alternator or power steering pump.

- A serpentine belt, which also drives the alternator, power steering pump and AC compressor among other things.

- The timing belt that is also responsible for driving one or more camshafts.

The water pump is made up of a housing, usually made of cast iron or cast aluminum and an impeller mounted on a spinning shaft with a pulley attached to the shaft on the outside of the pump body. A seal keeps fluid from leaking out of the pump housing past the spinning shaft. The impeller uses centrifugal force to draw the coolant in from the lower radiator hose and send it under pressure into the engine block. There is a gasket to seal the water pump to the engine block and prevent the flowing coolant from leaking out where the pump is attached to the block.

Thermostat

The thermostat is simply a valve that measures the temperature of the coolant and, if it is hot enough, opens to allow the coolant to flow through the radiator. If the coolant is not hot enough, the flow to the radiator is blocked and fluid is directed to a bypass system that allows the coolant to return directly back to the engine. The bypass system allows the coolant to keep moving through the engine to balance the temperature and avoid hot spots. Because flow to the radiator is blocked, the engine will reach operating temperature sooner and, on a cold day, will allow the heater to begin supplying hot air to the interior more quickly.

Since the 1970s, thermostats have been calibrated to keep the temperature of the coolant above 192 to 195 degrees. Prior to that, 180 degree thermostats were the norm. It was found that if the engine is allowed to run at these hotter temperatures, emissions are reduced, moisture condensation inside the engine is quickly burned off extending engine life, and combustion is more complete which improves fuel economy.

The heart of a thermostat is a sealed copper cup that contains wax and a metal pellet. As the thermostat heats up, the hot wax expands, pushing a piston against spring pressure to open the valve and allow coolant to circulate.

The thermostat is usually located in the front, top part of the engine in a water outlet housing that also serves as the connection point for the upper radiator hose. The thermostat housing attaches to the engine, usually with two bolts and a gasket to seal it against leaks. The gasket is usually made of a heavy paper or a rubber O ring is used. In some applications, there is no gasket or rubber seal. Instead, a thin bead of special silicone sealer is squeezed from a tube to form a seal.

There is a mistaken belief by some people that if they remove the thermostat, they will be able to solve hard to find overheating problems. This couldn't be further from the truth. Removing the thermostat will allow uncontrolled circulation of the coolant throughout the system. It is possible for the coolant to move so fast, that it will not be properly cooled as it races through the radiator, so the engine can run even hotter than before under certain conditions. Other times, the engine will never reach its operating temperature. On computer controlled vehicles, the computer monitors engine temperatures and regulates fuel usage based on that temperature. If the engine never reaches operating temperatures, fuel economy and performance will suffer considerably.

Bypass System

This is a passage that allows the coolant to bypass the radiator and return directly back to the engine. Some engines use a rubber hose, or a fixed steel tube. In other engines, there is a cast in passage built into the water pump or front housing. In any case, when the thermostat is closed, coolant is directed to this bypass and channeled back to the water pump, which sends the coolant back into the engine without being cooled by the radiator.

Freeze Plugs

When an engine block is manufactured, a special sand is molded to the shape of the coolant passages in the engine block. This sand sculpture is positioned inside a mold and molten iron or aluminum is poured to form the engine block. When the casting is cooled, the sand is loosened and removed through holes in the engine block casting leaving the passages that the coolant flows through. Obviously, if we don't plug up these holes, the coolant will pour right out.

Plugging these holes is the job of the freeze-out plug. These plugs are steel discs or cups that are press fit in the holes in the side of the engine block and normally last the life of the engine with no problems. But there is a reason they are called freeze-out plugs. In the early days, many people used plain water in their engines, usually after replacing a burst hose or other cooling system repair.

Needless to say, people are forgetful and many a motor suffered the fate of the water freezing inside the block. Often, when this happened the pressure of the water freezing and expanding forced the freeze-out plugs to pop out, relieving the pressure and saving the engine block from cracking. (although, just as often the engine cracked anyway). Another reason for these plugs to fail was the fact that they were made of steel and would easily rust through if the vehicle owner was careless about maintaining the cooling system. Antifreeze has rust inhibitors in the formula to prevent this from happening, but those chemicals would lose their effect after 3 years, which is why antifreeze needs to be changed periodically. The fact that some people left plain water in their engines greatly accelerated the rusting of these freeze plugs.

When a freeze plug becomes so rusty that it perforates, you have a coolant leak that must be repaired by replacing the rusted out freeze plug with a new one. This job ranges from fairly easy to extremely difficult depending on the location of the affected freeze plug. Freeze plugs are located on the sides of the engine, usually 3 or 4 per side. There are also freeze plugs on the back of the engine on some models and also on the heads.

As long as you are good about maintaining the cooling system, you need never worry about these plugs failing on modern vehicles.

Head Gaskets and Intake Manifold Gaskets

All internal combustion engines have an engine block and one or two cylinder heads. The mating surfaces where the block and head meet are machined flat for a close, precision fit, but no amount of careful machining will allow them to be completely water tight or be able to hold back combustion gases from escaping past the mating surfaces.

In order to seal the block to the heads, we use a head gasket. The head gasket has several things it needs to seal against. The main thing is the combustion pressure on each cylinder. Oil and coolant must easily flow between block and head and it is the job of the head gasket to keep these fluids from leaking out or into the combustion chamber, or each other for that matter.

A typical head gasket is usually made of soft sheet metal that is stamped with ridges that surround all leak points. When the head is placed on the block, the head gasket is sandwiched between them. Many bolts, called head bolts are screwed in and tightened down causing the head gasket to crush and form a tight seal between the block and head.

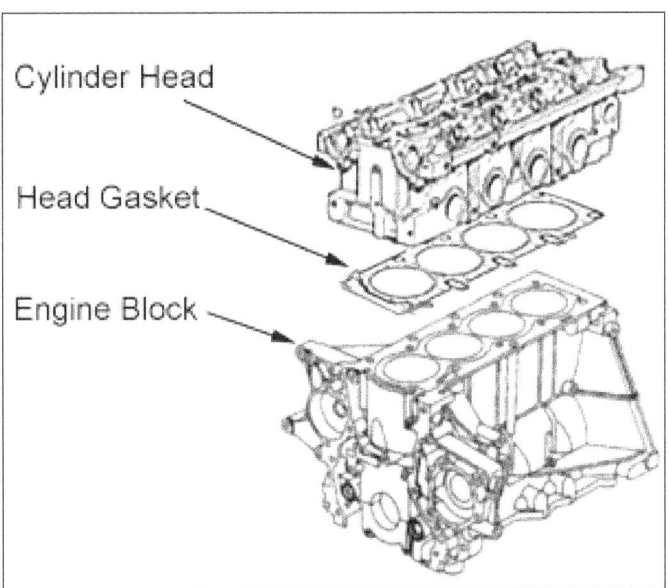

Cylinder Head

Head Gasket

Engine Block

Head gaskets usually fail if the engine overheats for a sustained period of time causing the cylinder head to warp and release pressure on the head gasket. This is most common on engines with cast aluminum heads, which are now on just about all modern engines.

Once coolant or combustion gases leak past the head gasket, the gasket material is usually damaged to a point where it will no longer hold the seal. This causes leaks in several possible areas. For example:

- Combustion gases could leak into the coolant passages causing excessive pressure in the cooling system.

- Coolant could leak into the combustion chamber causing coolant to escape through the exhaust system, often causing a white cloud of smoke at the tailpipe.

Other problems such as oil mixing with the coolant or being burned out the exhaust are also possible.

Some engines are more susceptible to head gasket failure than others.

Head gaskets themselves are relatively cheap, but it is the labor that's the killer. A typical head gasket replacement is a several hour job where the top part of the engine must be completely disassembled. These jobs can easily reach $1,000 or more.

On V type engines, there are two heads, meaning two head gaskets. While the labor won't double if both head gaskets need to be replaced, it will probably add a good 30% more labor to replace both. If only one head gasket has failed, it is usually not necessary to replace both, but it could be added insurance to get them both done at once.

A head gasket replacement begins with the diagnosis that the head gasket has failed. There is no way for a technician to know for certain whether there is additional damage to the cylinder head or other components without first disassembling the engine. All he or she knows is that fluid and combustion is not being contained.

One way to tell if a head gasket has failed is through a combustion leak test on the radiator. This is a chemical test that determines if there are combustion gases in the engine coolant. Another way is to remove the spark plugs and crank the engine while watching for water spray from one or more spark plug holes. Once the technician has determined that a head gasket must be replaced, an estimate is given for parts and labor. The technician will then explain that there may be additional charges after the engine is opened if more damage is found.

Heater Core

The hot coolant is also used to provide heat to the interior of the vehicle when needed. This is a simple and straight forward system that includes a heater core, which looks like a small version of a radiator, connected to the cooling system with a pair of rubber hoses. One hose brings hot coolant from the water pump to the heater core and the other hose returns the coolant to the top of the engine. There is usually a heater control valve in one of the hoses to block the flow of coolant into the heater core when maximum air conditioning is called for.

A fan, called a blower, draws air through the heater core and directs it through the heater ducts to the interior of the car. Temperature of the heat is regulated by a blend door that mixes cool outside air, or sometimes air conditioned air with the heated air coming through the heater core. This blend door allows you to control the temperature of the air coming into the interior. Other doors allow you to direct the warm air through the ducts on the floor, the defroster ducts at the base of the windshield, and the air conditioning ducts located in the instrument panel.

Hoses

There are several rubber hoses that make up the plumbing to connect the components of the cooling system. The main hoses are called the upper and lower radiator hoses. These two hoses are approximately 2 inches in diameter and direct coolant between the engine and the radiator. Two additional hoses, called heater hoses, supply hot coolant from the engine to the heater core. These hoses are approximately 1 inch in diameter. One of these hoses may have a heater control valve mounted in-line to block the hot coolant from entering the heater core when the air conditioner is set to max-cool. A fifth hose, called the bypass hose, is used to circulate the coolant through the engine, bypassing the radiator, when the thermostat is closed. Some engines do not use a rubber hose. Instead, they might use a metal tube or have a built-in passage in the front housing.

These hoses are designed to withstand the pressure inside the cooling system. Because of this, they are subject to wear and tear and eventually may require replacing as part of routine maintenance. If the rubber is beginning to look dry and cracked, or becomes soft and spongy, or you notice some ballooning at the ends, it is time to replace them. The main radiator hoses are usually molded to a shape that is designed to rout the hose around obstacles without kinking. When purchasing replacements, make sure that they are designed to fit the vehicle.

There is a small rubber hose that runs from the radiator neck to the reserve bottle. This allows coolant that is released by the pressure cap to be sent to the reserve tank. This rubber hose is about a quarter inch in diameter and is normally not part of the pressurized system. Once the engine is cool, the coolant is drawn back to the radiator by the same hose.

Cooling System Maintenance and Repair

An engine that is overheating will quickly self destruct, so proper maintenance of the cooling system is very important to the life of the engine and the trouble free operation of the cooling system in general.

The most important maintenance item is to flush and refill the coolant periodically. The reason for this important service is that anti-freeze has a number of additives that are designed to prevent corrosion in the cooling system. This corrosion tends to accelerate when several different types of metal interact with each other. The corrosion causes scale that eventually builds up and begins to clog the thin flat tubes in the radiator and heater core. causing the engine to eventually overheat. The anti-corrosion chemicals in the antifreeze prevents this, but they have a limited life span.

Newer antifreeze formulations will last for 5 years or 150,000 miles before requiring replacement. These antifreezes are usually red in color and are referred to as "Extended Life" or "Long Life" antifreeze. GM has been using this type of coolant in all their vehicles since 1996. The GM product is called "Dex-Cool".

Most antifreeze used in vehicles however, is green in color and should be replaced every two years or 30,000 miles, which ever comes first. You can convert to the new long life coolant, but only if you completely flush out all of the old antifreeze. If any green coolant is allowed to mix with the red coolant, you must revert to the shorter replacement cycle.

Look for a shop that can reverse-flush the cooling system. This requires special equipment and the removal of the thermostat in order to do the job properly. This type of flush is especially important if the old coolant looks brown or has scale or debris floating around in it.

If you remove the thermostat for a reverse flush, always replace it with a new thermostat of the proper temperature. It is cheap insurance.

The National Automotive Radiator Service Association (NARSA) recommends that motorists have a seven-point preventative cooling system maintenance check at least once every two years. The seven-point program is designed to identify any areas that need attention. It consists of:

- A visual inspection of all cooling system components, including belts and hoses.

- A radiator pressure cap test to check for the recommended system pressure level.

- A thermostat check for proper opening and closing.

- A pressure test to identify any external leaks to the cooling system parts; including the radiator, water pump, engine coolant passages, radiator and heater hoses and heater core.

- An internal leak test to check for combustion gas leakage into the cooling system.

- An engine fan test for proper operation.

- A system power flush and refill with car manufacturer's recommended concentration of coolant.

Let's take these items one at a time.

Visual Inspection

What you are looking for is the condition of the belts and hoses. The radiator hoses and heater hoses are easily inspected just by opening the hood and looking. You want to be sure that the hoses have no cracking or splitting and that there is no bulging or swelling at the ends. If there is any sign of problems, the hose should be replaced with the correct part number for the year, make and model of the vehicle. Never use a universal hose unless it is an emergency and a proper molded hose is not available.

Heater hoses are usually straight runs and are not molded, so a universal hose is fine to use and often is all that is available. Make sure that you use the proper inside diameter for the hose being replaced. For either the radiator hoses or the heater hoses, make sure that you route the replacement hose in the same way that the original hose was running. Position the hose away from any obstruction that can possibly damage it and always use new hose clamps. After you refill the cooling system with coolant, do a pressure test to make sure that there are no leaks.

On most older vehicles, the water pump is driven by a V belt or serpentine belt on the front of the engine that is also responsible for driving the alternator, power steering pump and air conditioner compressor. These types of belts are easy to inspect and replace if they are worn. You are looking for dry cracking on the inside surface of the belt.

On later vehicles, the water pump is often driven by the timing belt. This belt usually has a specific life expectancy at which time it must be replaced to insure that it does not fail. Since the timing belt is inside the engine and will require partial engine disassembly to inspect, it is very important to replace it at the correct interval. Since the labor to replace this belt can be significant, it is a good idea to replace the water pump at the same time that the belt is replaced. This is because 90 percent of the labor to replace a water pump has already been done to replace the timing belt. It is simply good insurance to replace the pump while everything is apart.

Radiator Pressure Cap Test

A radiator pressure cap is designed to maintain pressure in the cooling system at a certain maximum pressure. If the cooling system exceeds that pressure, a valve in the cap opens to bleed the excessive pressure into the reserve tank. Once the engine has cooled off, a negative pressure begins to develop in the cooling system. When this happens, a second valve in the cap allows the coolant to be siphoned back into the radiator from the reserve tank. If the cap should fail, the engine can easily overheat. A pressure test of the radiator cap is a quick way to tell if the cap is doing its job. It should be able to hold its rated pressure for two minutes.

Thermostat Check for Proper Opening and Closing

This step is only necessary if you are having problems with the cooling system. A thermostat is designed to open at a certain coolant temperature. To test a thermostat while it is still in the engine, start the engine and let it come to normal operating temperature (do not let it overheat). If it takes an unusually long time for the engine to warm up or for the heater to begin delivering hot air, the thermostat may be stuck in the open position. If the engine does warm up, shut it off and look for the two radiator hoses. These are the two large hoses that go from the engine to the radiator. Feel them carefully (they could be very hot). If one hose is hot and the other is cold, the thermostat may be stuck closed.

If you are having problems and suspect the thermostat, remove it and place it in a pot of water. Bring the water to a boil and watch the thermostat. You should see it open when the water reaches a boil. Most thermostats open at about 195 degrees Fahrenheit. An oven thermometer in the water should confirm that the thermostat is working properly.

Pressure Test to Identify any External Leaks

Pressure testing the cooling system is a simple process to determine where a leak is located. This test is only performed after the cooling system has cooled sufficiently to allow you to safely remove the pressure cap. Once you are sure that the cooling system is full of coolant, a cooling system pressure tester is attached in place of the radiator cap. The tester is than pumped to build up pressure in the system. There is a gauge on the tester indicating how much pressure is being pumped. You should pump it to the pressure indicated on the pressure cap or to manufacturer's specs.

Once pressure is applied, you can begin to look for leaks. Also watch the gauge on the tester to see if it loses pressure. If the pressure drops more than a couple of pounds in two minutes, there is likely a leak somewhere that may be hidden. It is not always easy to see where a leak is originating from. It is best to have the vehicle up on a lift so you can look over everything with a shop light or flashlight. If the heater core in leaking, it may not be visible since the core is enclosed and not visible without major disassembly, but one sure sign is the unmistakable odor of antifreeze inside the car. You may also notice the windshield steaming up with an oily residue.

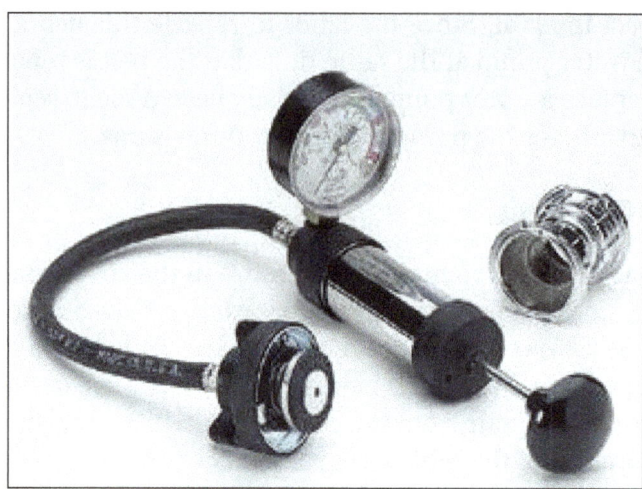
Cooling system pressure tester.

Internal Leak Test

If you are losing coolant, but there are no signs of leaks, you could have a blown head gasket. The best way to test for this problem is with a combustion leak test on the radiator. This is accomplished using a block tester. This is a kit that performs a chemical test on the vapors in the radiator. Blue tester fluid is added to the plastic container on the tester. If the fluid turns yellow during the test, then exhaust gasses are present in the radiator.

The most common causes for exhaust gasses to be present in the radiator is a blown head gasket. Replacing a bad head gasket requires a major disassembly of the engine and can be quite expensive. Other causes include a cracked head or a cracked block, both are even more undesirable than having to replace a head gasket.

When a Head Gasket goes Bad

The process of replacing a head gasket begins with completely draining the coolant from the engine. The top part of the engine is then disassembled along with much of the front of the engine in order to gain access to the cylinder heads. The head or heads are then removed and a thorough inspection for additional damage is done.

Before the engine can be reassembled, the mating surfaces of the head and block are first cleaned to make sure that nothing will interfere with the sealing properties of the gasket. The surface of the cylinder head is also checked for flatness and, in some cases, the block is checked as well. The head gasket is then positioned on the block and aligned using locator pegs that are built into the block. The head is then placed on top of the gasket and a number of bolts, called head-bolts are coated with oil and loosely threaded into the assembly. The bolts are then tightened in a specific order to a specified initial torque using a special wrench called a torque wrench. This is to insure that the head gasket is crushed evenly in order to insure a tight seal. This process is then repeated to a second, tighter torque setting, then finally a third torque setting. At this point, the rest of the engine is reassembled and the cooling system is filled with a mixture of antifreeze and water. Once the engine is filled, the technician will pressure test the cooling system to make sure there are no leaks.

In many engines, coolant also passes between the heads and the intake manifold. There are also gaskets for the intake manifold to keep the coolant from leaking out at that point. Replacing an intake manifold gasket is a much easier job than a head gasket, but can still take a couple of hours or more for that job.

Engine Fan Test

The radiator cooling fan is an important part of the cooling system operation. While a fan is not really needed while a vehicle is traveling down the highway, it is extremely important when driving slowly or stopped with the engine running. In the past, the fan was attached to the engine and was driven by the fan belt. The speed of the fan was directly proportional to the speed of the engine. This type of system sometimes caused excessive noise as the car accelerated through the gears. As the engine sped up, a rushing fan noise could be heard. To quiet things down and place less of a drag in the engine, a viscous fan drive was developed in order to disengage the fan when it was not needed.

When computer controls came into being, these engine driven fans gave way to electric fans that were mounted directly on the radiator. A temperature sensor determined when the engine was beginning to run too hot and turned on the fan to draw air through the radiator to cool the engine. On many cars, there were two fans mounted side by side to make sure that the radiator had a uniform air flow for the width of the unit.

When the car was in motion, the speed of the air entering the grill was sufficient to keep the coolant at the proper temperature, so the fans were shut off. When the vehicle came to a stop, there was no natural air flow, so the fan would come on as soon as the engine reached a certain temperature.

If the air conditioner was turned on, a different circuit would come into play. The reason for this is the air conditioning system always requires a good air flow through the condenser mounted in front of the radiator. If the air flow stopped, the air conditioned air coming through the dash outlets would immediately start warming up. For this reason, when the air conditioner is turned on, the fan circuit would power the fans regardless of engine temperature.

If you notice that the engine temperature begins rising soon after the vehicle comes to a stop, the first thing to check is fan operation. If the fan is not turning when the engine is hot, a simple test is to turn the AC on. If the fan begins to work, suspect the temperature sensor in the fan circuit. In order to test the fan motor itself, unplug the two wire connector to the fan and connect a 12 volt source to one terminal and ground the other. (it doesn't matter which is which for this test) If the fan motor begins to turn, the motor is good. If it doesn't turn, the motor is bad and must be replaced.

In order to test the system further, you will need a repair manual for the year, make and model vehicle and follow the troubleshooting charts and diagnostic procedures for your vehicle. On most systems, there will be a fan relay or fan control module that can be a trouble spot. There are a number of different control systems, each requiring a different test procedure. Without the proper repair information, you can easily do more harm than good.

Cooling System Power Flush and Refill

While you can replace old coolant by draining it out and replacing it with fresh coolant, the best way to properly maintain your cooling system is to have the system power flushed. Power flushing will remove all the old coolant and pull out any sediment and scale along with it.

Power flushing requires a special machine that many auto repair shops have for the purpose. The procedure requires that the thermostat is removed, the lower radiator hose is disconnected, and the flush machine is connected in line. The lower hose is connected to the machine and the other hose from the machine is connected to the radiator where the lower hose was disconnected from.

Water, and sometimes, a cleaning agent is pumped through the cooling system in a reverse path from the normal coolant flow. This allows any scale to be loosened and flow out. Once clear water is coming out of the system, the hose is reconnected and a new thermostat is installed. Then the cooling system is refilled with the appropriate amount of antifreeze to bring the coolant to the proper mixture of antifreeze and water. For most vehicles and most climates, the mixture is 50 percent antifreeze and 50 percent water. In colder climates, more antifreeze is used, but must never exceed 75 percent antifreeze.

ELECTRICAL SYSTEM

The electrical system comprises a storage battery, generator, starting (cranking) motor, lighting system, ignition system, and various accessories and controls. Originally, the electrical system of the automobile was limited to the ignition equipment. With the advent of the electric starter on a 1912 Cadillac model, electric lights and horns began to replace the kerosene and acetylene lights and the bulb horns. Electrification was rapid and complete, and, by 1930, 6-volt systems were standard everywhere.

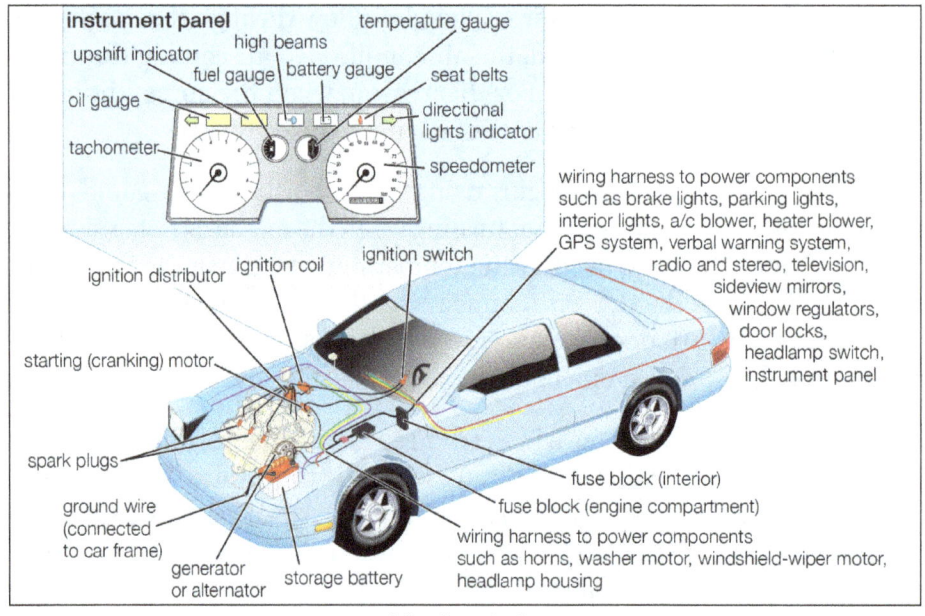

Automobile electrical systems.

Increased engine speeds and higher cylinder pressures made it increasingly difficult to meet high ignition voltage requirements. The larger engines required higher cranking torque. Additional electrically operated features—such as radios, window regulators, and multispeed windshield wipers—also added to system requirements. To meet these needs, 12-volt systems replaced the 6-volt systems in the late 1950s around the world.

The ignition system provides the spark to ignite the air-fuel mixture in the cylinders of the engine. The system consists of the spark plugs, coil, distributor, and battery. In order to jump the gap between the electrodes of the spark plugs, the 12-volt potential of the electrical system must be stepped up to about 20,000 volts. This is done by a circuit that starts with the battery, one side of which is grounded on the chassis and leads through the ignition switch to the primary winding of the ignition coil and back to the ground through an interrupter switch. Interrupting the primary circuit induces a high voltage across the secondary terminal of the coil. The high-voltage secondary terminal of the coil leads to a distributor that acts as a rotary switch, alternately connecting the coil to each of the wires leading to the spark plugs.

Solid-state or transistorized ignition systems were introduced in the 1970s. These distributor systems provided increased durability by eliminating the frictional contacts between breaker points and distributor cams. The breaker point was replaced by a revolving magnetic-pulse generator in which alternating-current pulses trigger the high voltage needed for ignition by means of an amplifier electronic circuit. Changes in engine ignition timing are made by vacuum or electronic control unit (microprocessor) connections to the distributor.

The source of energy for the various electrical devices of the automobile is a generator, or alternator, that is belt-driven from the engine crankshaft. The design is usually an alternating-current type with built-in rectifiers and a voltage regulator to match the generator output to the electric load and also to the charging requirements of the battery, regardless of engine speed.

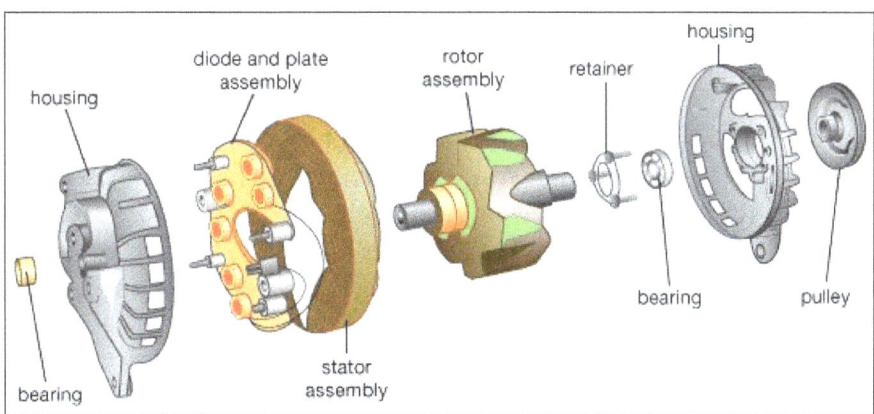

Automobile Alternator.

In the above figure is shown exploded view of an automotive alternator. The engine's turning crankshaft, connected to the alternator's pulley by a belt, turns the magnetic rotor inside the stationary stator assembly, generating an alternating current. The diode assembly rectifies the alternating current, producing direct current, which is used to meet the demands of the vehicle's electrical system, including recharging the battery.

A lead-acid battery serves as a reservoir to store excess output of the generator. This provides energy for the starting motor and power for operating other electric devices when the engine is not running or when the generator speed is not sufficiently high for the load.

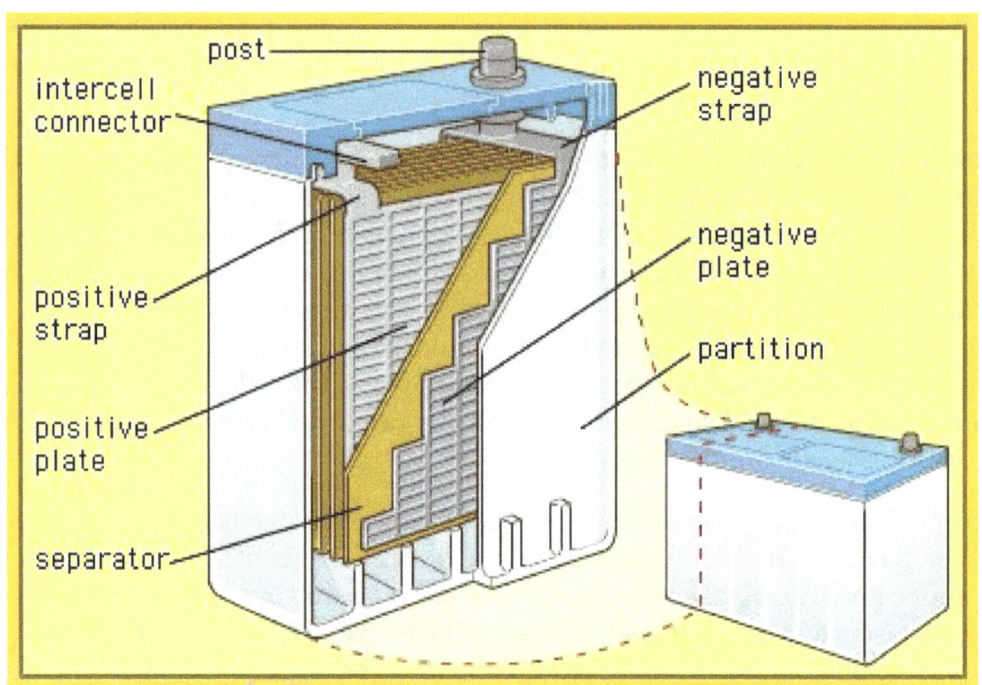

Construction of the automotive-type lead-acid battery (cutaway view). A storage battery not only holds its charge for a long time, but it also can be recharged.

The starting motor drives a small spur gear so arranged that it automatically moves into mesh with gear teeth on the rim of the flywheel as the starting-motor armature begins to turn. When the engine starts, the gear is disengaged, thus preventing damage to the starting motor from over-speeding. The starting motor is designed for high current consumption and delivers considerable power for its size for a limited time.

Headlights must satisfactorily illuminate the highway ahead of the automobile for driving at night or in inclement weather without temporarily blinding approaching drivers. This was achieved in modern cars with double-filament bulbs with a high and a low beam, called sealed-beam units. Such units could have only one filament at the focal point of the reflector. Because of the greater illumination required for high-speed driving with the high beam, the lower beam filament was placed off centre, with a resulting decrease in lighting effectiveness. Separate lamps for these functions can also be used to improve illumination effectiveness.

Dimming is automatically achieved on some cars by means of a photocell-controlled switch in the lamp circuit that is triggered by the lights of an oncoming car. Lamp clusters behind aerodynamic plastic covers permitted significant front-end drag reduction and improved fuel economy. In this arrangement, steerable headlights became possible with an electric motor to swivel the lamp assembly in response to steering wheel position. The regulations of various governments dictate brightness and field of view requirements for vehicle lights.

Signal lamps and other special-purpose lights have increased in usage since the 1960s. Amber-coloured front and red rear signal lights are flashed as a turn indication; all these lights are flashed simultaneously in the "flasher" (hazard) system for use when a car is parked along a roadway or is traveling at a low speed on a high-speed highway. Marker lights that are visible from the front, side, and rear also are widely required by law. Red-coloured rear signals are used to denote braking, and cornering lamps, in connection with turning, provide extra illumination in the direction of an intended turn. Backup lights provide illumination to the rear and warn anyone behind the vehicle when the driver is backing up. High-voltage light-emitting diodes (LEDs) are under development for various signal and lighting applications.

TRANSMISSION SYSTEM

The transmission systems in vehicles are installed to send signals from one part to another such as from engines to wheels to ensure mobility. In this process, several components that are fixed in the back of the engine in a car play integral part. There is limited variety we find in the transmission devices across the globe. However, some of the essential components of transmission system are modulator, torque converter, planetary gears, governor, computer, seals, output shafts, differential, hydraulic designs, axles an pressure plate.

Types of Transmission System

The automobile transmission system can be classified into two types including manual transmission and automatic transmission. In case of the manual transmission system, the vehicle is driven with the assistance of gearshift and foot clutch. The other components, which are used in this process, are flywheel, pressure plate and ring gears.

In case of the automatic transmission system, the gears are changed automatically corresponding with the vehicle's speed. The basic components essential for this process are modulator, torque converter, planetary gears, governor, computer, seals and hydraulic designs.

Features of Transmission Systems

The manual as well as automatic transmission systems have their own advantages and disadvantages. To start with, the manual transmission system increases fuel efficiency of the vehicles. It is cost-effective in terms of maintenance. In this process, the driver has the optimum control over the vehicle. Automobiles enabled with manual transmission system can overtake other vehicles much easily as it comes with extra throttle valve. It allows driver to go for lower gear that makes the process more powerful.

The automatic transmission systems remove the task of depressing clutch pedal and choosing a gear ratio. The process makes the task simple and easy. Unlike manual transmission, the auto transmission system has only two pedals instead of three. To select any mode, the diver needs to use a gear-shift lever fixed on the car floor. Here, the person needs to maneuver through shift lock buttons provided on the lever.

In the vehicles enabled with automatic transmission system, you would find P for Park, R for Reverse, N for Neutral and D for Drive. On the selection of option P, the transmission system gets locked mechanically. The car stops to move. The Neutral option is selected to stop the wheels completely. If driver puts car on Neutral mode, one cannot start it. However, car can be shifted forth or back smoothly. Reverse mode helps the driver to drive in reverse direction. For this, the driver needs to push the shift lock button on the lever. Drive option allows the car to go in forward direction through various forward gears.

Some of the cars in India, which have set the best examples of the automatic transmission system, are Skoda Superb, Toyota Camry and Honda CR-V. Nevertheless, both the systems have certain disadvantages also. In manual transmissions, the driver needs to be very careful and proficient. It loses to automatic transmission system in case of driving in the hilly regions. As per automatic transmission system is concerned, it does not allow optimum fuel efficiency.

Tips for Buying Transmission System Parts

There are few things you need to keep in mind while planning to buy any automobile transmission system parts. These are:

- Look for fresh automobile parts.

- Look for certified suppliers.

- Look for the product specific info.

- Look into the terms and conditions of the products before buying.

OTHER MECHANICAL SUBSYSTEMS

The Braking System

Most modern cars have brakes on all four wheels, operated by a hydraulic system . The brakes may be disc type or drum type.

The front brakes play a greater part in stopping the car than the rear ones, because braking throws the car weight forward on to the front wheels.

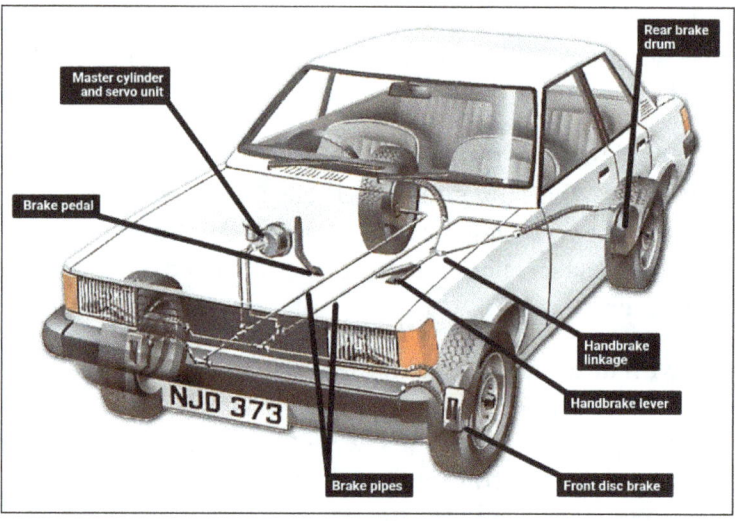

Dual-circuit Braking System.

A typical dual-circuit braking system in which each circuit acts on both front wheels and one rear wheel. Pressing the brake pedal forces fluid out of the master cylinder along the brake pipes to the slave cylinders at the wheels; the master cylinder has a reservoir that keeps it full.

Many cars therefore have disc brakes, which are generally more efficient, at the front and drum brakes at the rear.

All-disc braking systems are used on some expensive or high-performance cars, and all-drum systems on some older or smaller cars.

Brake Hydraulics

A hydraulic brake circuit has fluid-filled master and slave cylinders connected by pipes.

When you push the brake pedal it depresses a piston in the master cylinder, forcing fluid along the pipe.

The fluid travels to slave cylinders at each wheel and fills them, forcing pistons out to apply the brakes.

Fluid pressure distributes itself evenly around the system.

The combined surface 'pushing' area of all the slave pistons is much greater than that of the piston in the master cylinder.

Consequently, the master piston has to travel several inches to move the slave pistons the fraction of an inch it takes to apply the brakes.

This arrangement allows great force to be exerted by the brakes, in the same way that a long-handled lever can easily lift a heavy object a short distance.

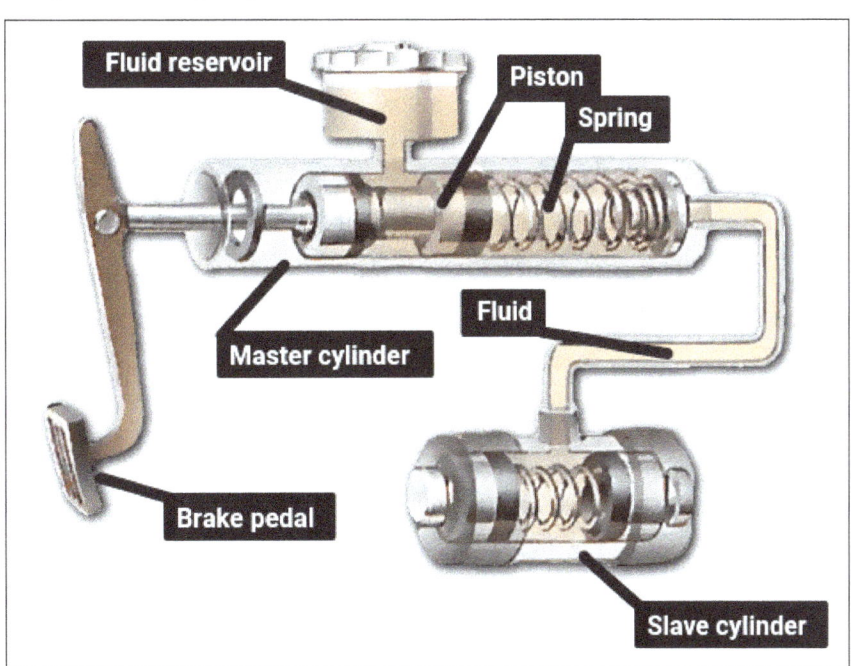

Master and Slave Cylinders The master cylinder transmits hydraulic
pressure to the slave cylinder when the pedal is pressed.

Most modern cars are fitted with twin hydraulic circuits, with two master cylinders in tandem, in case one should fail.

Sometimes one circuit works the front brakes and one the rear brakes; or each circuit works both front brakes and one of the rear brakes; or one circuit works all four brakes and the other the front ones only.

Under heavy braking, so much weight may come off the rear wheels that they lock, possibly causing a dangerous skid.

For this reason, the rear brakes are deliberately made less powerful than the front.

Most cars now also have a load-sensitive pressure-limiting valve. It closes when heavy braking raises hydraulic pressure to a level that might cause the rear brakes to lock, and prevents any further movement of fluid to them.

Advanced cars may even have complex anti-lock systems that sense in various ways how the car is decelerating and whether any wheels are locking.

Such systems apply and release the brakes in rapid succession to stop them locking.

Power-assisted Brakes

Many cars also have power assistance to reduce the effort needed to apply the brakes.

Usually the source of power is the pressure difference between the partial vacuum in the inlet manifold and the outside air.

The servo unit that provides the assistance has a pipe connection to the inlet manifold.

A direct-acting servo is fitted between the brake pedal and the master cylinder. The brake pedal pushes a rod that in turn pushes the master-cylinder piston.

But the brake pedal also works on a set of air valves, and there is a large rubber diaphragm connected to the master-cylinder piston.

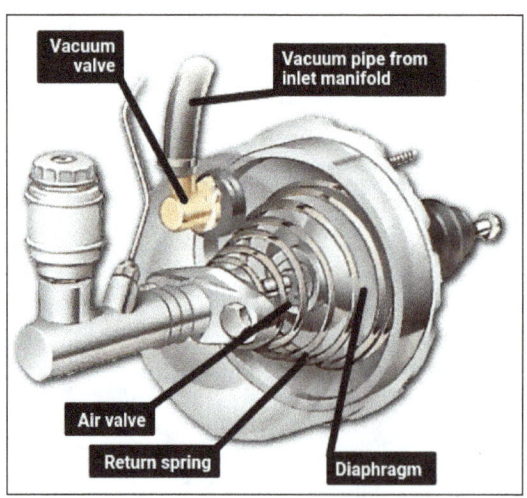

A direct-acting servo is fitted between the brake pedal and the master cylinder. The pedal can work the master cylinder directly if the servo fails or if the engine is not running.

When the brakes are off, both sides of the diaphragm are exposed to the vacuum from the manifold.

Pressing the brake pedal closes the valve linking the rear side of the diaphragm to the manifold, and opens a valve that lets in air from outside.

The higher pressure of the outside air forces the diaphragm forward to push on the master-cylinder piston, and thereby assists the braking effort.

If the pedal is then held, and pressed no further, the air valve admits no more air from outside, so the pressure on the brakes remains the same.

When the pedal is released, the space behind the diaphragm is reopened to the manifold, so the pressure drops and the diaphragm falls back.

If the vacuum fails because the engine stops, for example the brakes still work because there is a normal mechanical link between the pedal and the master cylinder. But much more force must be exerted on the brake pedal to apply them.

Some cars have an indirect-acting servo fitted in the hydraulic lines between the master cylinder and the brakes. Such a unit can be mounted anywhere in the engine compartment instead of having to be directly in front of the pedal.

It, too, relies on manifold vacuum to provide the boost. Pressing the brake pedal causes hydraulic pressure build up from the master cylinder, a valve opens and that triggers the vacuum servo.

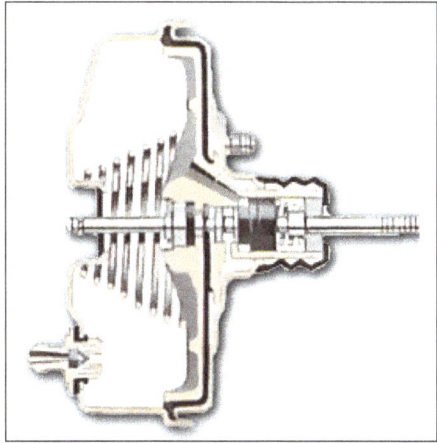

Brake off – both sides of the diaphragm are under vacuum.

Disc Brakes

A disc brake has a disc that turns with the wheel. The disc is straddled by a caliper , in which there are small hydraulic pistons worked by pressure from the master cylinder.

The pistons press on friction pads that clamp against the disc from each side to slow or stop it. The pads are shaped to cover a broad sector of the disc.

There may be more than a single pair of pistons, especially in dual-circuit brakes.

The pistons move only a tiny distance to apply the brakes, and the pads barely clear the disc when the brakes are released. They have no return springs.

When the brake is applied, fluid pressure forces the pads against the disc. With the brake off, both pads barely clear the disc.

Rubber sealing rings round the pistons are designed to let the pistons slip forward gradually as the pads wear down, so that the tiny gap remains constant and the brakes do not need adjustment.

Many later cars have wear sensors leads embedded in the pads. When the pads are nearly worn out, the leads are exposed and short-circuited by the metal disc, illuminating a warning light on the instrument panel.

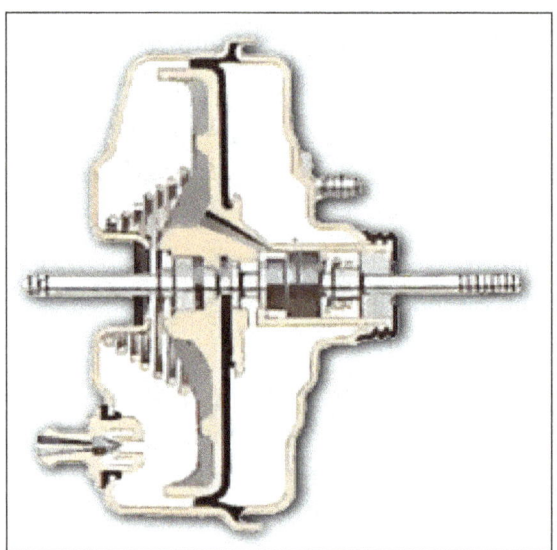

Applying the brake lets air in behind the diaphragm, forcing it against the cylinder.

Drum Brakes

A drum brake has a hollow drum that turns with the wheel. Its open back is covered by a stationary backplate on which there are two curved shoes carrying friction linings.

The shoes are forced outwards by hydraulic pressure moving pistons in the brake's wheel cylinders, so pressing the linings against the inside of the drum to slow or stop it.

Each brake shoe has a pivot at one end and a piston at the other. A leading shoe has the piston at the leading edge relative to the direction in which the drum turns.

The rotation of the drum tends to pull the leading shoe firmly against it when it makes contact, improving the braking effect.

Some drums have twin leading shoes, each with its own hydraulic cylinder; others have one leading and one trailing shoe – with the pivot at the front.

The basic type of disc brake, with a single pair of pistons. There may be more than one pair, or a single piston operating both pads, like a scissor mechanism, through different types of calipers – a swinging or a sliding caliper.

This design allows the two shoes to be forced apart from each other by a single cylinder with a piston in each end.

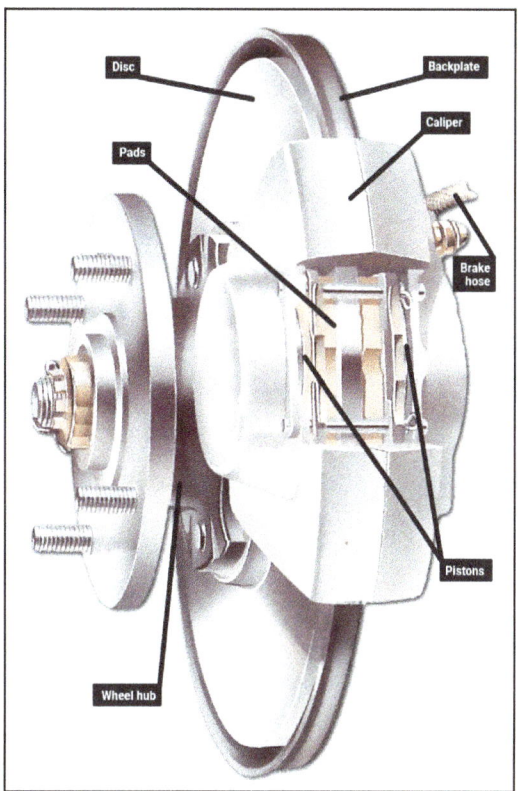

Disc brake.

It is simpler but less powerful than the two-leading-shoe system, and is usually restricted to rear brakes.

In either type, return springs pull the shoes back a short way when the brakes are released.

Shoe travel is kept as short as possible by an adjuster. Older systems have manual adjusters that need to be turned from time to time as the friction linings wear. Later brakes have automatic adjustment by means of a ratchet.

Drum brakes may fade if they are applied repeatedly within a short time - they heat up and lose their efficiency until they cool down again. Discs, with their more open construction, are much less prone to fading.

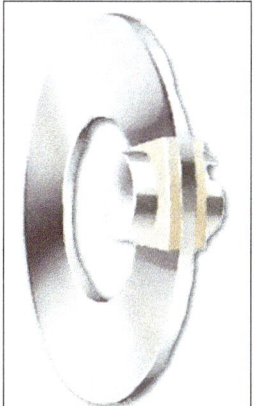

When the brake is applied, fluid pressure forces the pads against
the disc. With the brake off, both pads barely clear the disc.

The Handbrake

Apart from the hydraulic braking system, all cars have a mechanical handbrake acting on two wheels – usually the rear ones.

The handbrake gives limited braking if the hydraulic system fails completely, but its main purpose is as a parking brake .

The handbrake lever pulls a cable or pair of cables linked to the brakes by a set of smaller levers, pulleys and guides whose details vary greatly from car to car.

A ratchet on the handbrake lever keeps the brake on once it is applied. A push button disengages the ratchet and frees the lever.

On drum brakes, the handbrake system presses the brake linings against the drums.

With the brakes on, the shoes are forced against the drums by their piston.

The handbrake mechanism.

The handbrake acts on the shoes by means of a mechanical system, separate from the hydraulic cylinder, consisting of a lever and arm in the brake drum; they are operated by a cable from the handbrake lever inside the car.

Front Axle System

Front axle carries the weight of the front part of the automobile as well as facilitates steering and absorbs shocks due to road surface variations. The front axles are generally dead axles, but are live axles in small cars of compact designs and also in case of four-wheel drive. The steering system converts the rotary motion of the driver's steering wheel into the angular turning of the front wheels as well as to multiply the driver's effort with leverage or mechanical advantage for turning the wheels. The steering system, in addition to directing the vehicle in a particular direction must be arranged geometrically in such a way so that the wheels undergo true rolling motion without slipping or scuffing. Moreover, the steering must be light and stable with a certain degree of self-adjusting ability. Steering systems may also be power assisted.

Front Axle

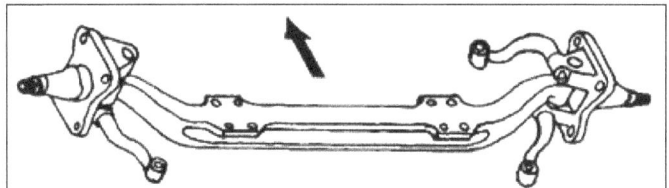

Front axle.

The front axle is designed to transmit the weight of the automobile from the springs to the front wheels, turning right or left as required. To prevent interference due to front engine location, and for providing greater stability and safety at high speeds by lowering the centre of gravity of the road vehicles, the entire centre portion of the axle is dropped. As shown in Fig. front axle includes the axle-beam, stub-axles with brake assemblies, u ack-rod and stub-axle arm. Front axles can be live axles and dead axles. A live front axle contains the differential mechanism through which the engine power flows towards the front wheels. For steering the front wheels, constant velocity joints are contained in the axle half shafts. Without affecting the power flow through the half shafts, these joints help in turning the stub axles around the king-pin. The front axles are generally dead axles, which does not transmit power. The front wheel hubs rotate on antifriction bearings of tapered-roller type on the steering spindles, which are an integral part of steering knuckles. To permit the wheels to be turned by the steering gear, the steering spindle and steering knuckle assemblies are hinged on the end of axle. The pin that forms the pivot of this hinge is known as king pin or steering knuckle pin. Generally dead front axles are three types. In the Elliot type front axles the yoke for king spindle is located on the ends of I-beam. The axle ends are forked to hold the steering knuckle extension between them. The reverse Elliot front axles have hinged spindle yoke on spindle itself instead of on the axle. The forked portion is integral with the steering knuckle. This type is commonly used as this facilitates the mounting of brake backing plate on the forged legs of the steering knuckle. In the Lemoine type front axle, instead of a yoke type hinge, an L-shaped spindle is used which is attached to the end of the axle by means of a pivot. It is normally used in tractors.

The axle beam in use is of I or H-section and is manufactured from alloy forged steel for rigidity and strength. As compared to dead front axles, a totally different type of swivelling mechanism is used on the live front axle. To connect the wheel hub axles with driving axle shafts, constant velocity joints are used for the vehicles fitted with the front live axles. Tracta, Rzeppa (or Sheppa) on Bendix constant velocity or universal joints are normally used.

Front axles are subjected to both bending and shear stresses. In the static condition, the axle may be considered as a beam supported vertically upward at the ends i.e. at the centre of the wheels and loaded vertically downward at the centres of the spring pads. The vertical bending moment thus caused is zero at the point of support and rises linearly to a maximum at the point of loading and then remains constant.

Thus the maximum bending moment = Wl, Nm

where, W = The load on one wheel, N

I = The distance between the centre of wheel and the spring pad, m. Under dynamic conditions, the vertical bending moment is increased due to road roughness.

But its estimate is difficult and hence is generally accounted for through a factor of safety. The front axle also experiences a horizontal bending moment because of resistance to motion and this is of a nature similar to the vertical one but of very small magnitude and hence can be neglected except in those situations when it is comparatively large.

The resistance to motion also causes a torque in the case of drop type front axle as shown in Fig. Thus the portions projected after the spring pads are subjected to combined bending and torsion.

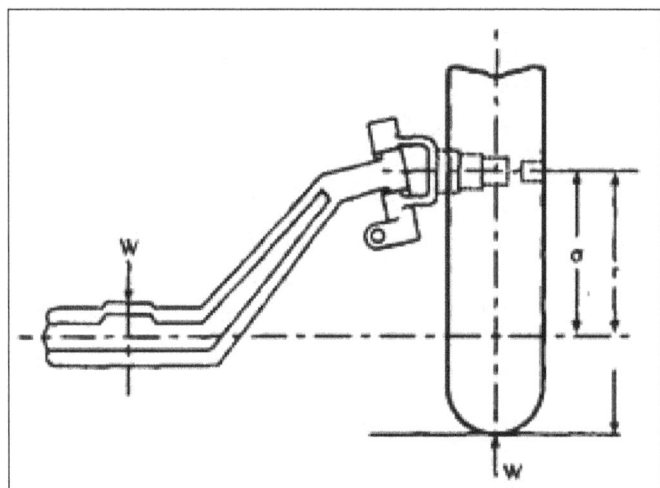

The magnitude of the torque= $R\delta$, Nm.

Where, R = the resistance to motion, N

δ =The drop from the spindle axis to the centre of the section, m

The shear stress in the axle is due to braking torque and its magnitude (as shown in Fig.)

$$= \mu\,W\,r, \text{Nm}.$$
r = The road wheel radius, m
μ = coefficient of adhesion between road and tyre
$= 0.6$ for dry, hard road surface(the maximum value of μ).

The braking torque is lower for the section lying between the spring pads and is given by $\mu W(r-\delta)$. In this portion the bending moment predominates whereas at the steering head, torsion predominates. Thus I-section in used for the portion where bending moment predominates and is gradually changed to circular, oval or rectangular section at the steering head.

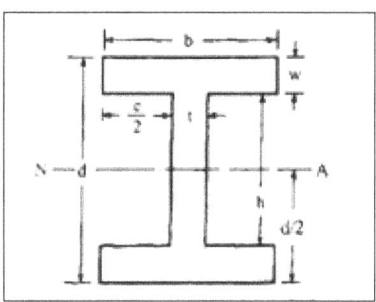

I-section, on the axle beam.

For I-section, the maximum bending moment is given by the relation,

$$\frac{M}{I} = \frac{f_b}{y}$$

where, M = the maximum bending moment, Nm

f_b = allow able bending stress for the material, N/m² or Pa

y = the maximum distance of the fibre from neutral axis (NA)

= $d/2$, m

I = The moment of inertia of the I- section about NA

= $\dfrac{bd^3 - ch^3}{12}$, m⁴ for I- section shown in fig

where, d = the over all depth of I- section, m

b = the flange width, m

t = the flange thickness, m

w = the web thickness, m

$c = b - t$

$h = d - 2w$

Generally, $d = 6t$

$b = 4.25t$

$w = 2.5t$

For the circular or oval section, the maximum torsion is given by the relation,

$$\frac{T}{Ip} = \frac{f_s}{y}$$

where, T = the maximum torque in the plane section, Nm

f_s = allowable shear stress in the material, N/m² or Pa

y = the distance from the neutral axis to the outermost fibre of the axle.

= $d/2$, m

d = the diameter for the ciruclar section

= the major axis for the oval section

I_p = Polar moment of intertia of the section

= $\dfrac{\pi}{32} d^4$ for circular section

= $\dfrac{\pi}{32} d^3 b$ for oval section with minor axis, b.

Types of Front Hub Assembly

The stub axle construction depends on whether it is a driving or non-driving hub.

Non-driving Hub

Figure A illustrates a typical bearing arrangement for a non-driving hub. This consists of a stub-axle, an externally cylindrical sleeve hub, a pair of taper-roller bearings, a grease-seal, a castellated adjustment nut and split-pin, a washer, and a dust-cap. A centrally flanged cylindrical sleeve hub is fitted over small outer and large inner taper-roller bearings, which are supported on the stub axle. The hub is made of malleable iron or steel cast. The bearings are designed to absorb both radial and axial loads when assembled. The slackness between the taper rollers and the inner and outer races are taken up by spinning the hub assembly while at the same time tightening the adjustment nut until all the free lay has been taken up. The bearings are then preloaded by tightening the nut with a torque wrench to some predetermined torque setting. The nut is then slackened slightly until one of the slots in the nut aligns with the hole in the stub-axle. The split pin is then inserted through and bent over to secure the nut in position.

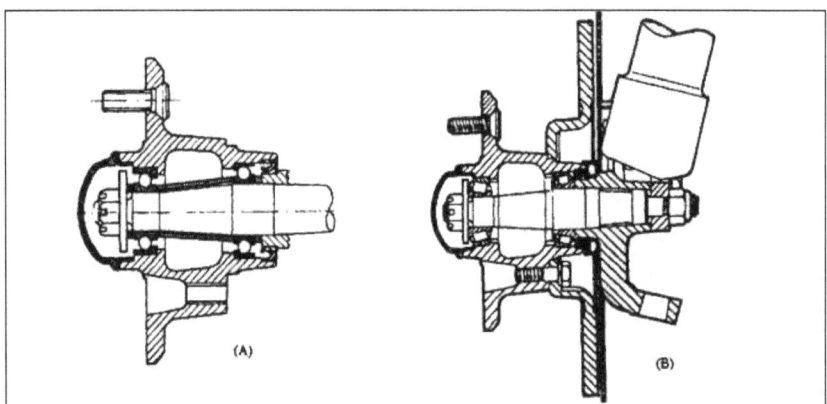

Front bearing-hub assembly. A. Taper roller type. B. Ball bearing type.

For more accurate setting, some manufactures recommend to tighten the adjusting nut to a specified torque before releasing the nut a given amount. This operation, however, should not be confused with preloading. As hub bearing withstands heat from the brakes, a running clearance is provided, which is quite different to that obtained after a bearing has been preloaded. The flanged hub supports the road wheels and a brake drum or a brake disc. A dust cap with a central vent hole is used to enclose the end of the hub. This area of the hub sliouid not be filled with grease.

An alternative bearing arrangement is presented in Fig. which uses two angular contact type ball races held apart by a rigid spacer. The nut on this non-adjustable hub are tightened fully to the correct torque value.

Driving Hub

Figure illustrates a typical hub arrangement used for a front-wheel drive car. The stub axle housing uses two bearings, which support both the wheel hub and driving shaft. The type of bearing used is decided based on the load carrying capacity of the road wheel.

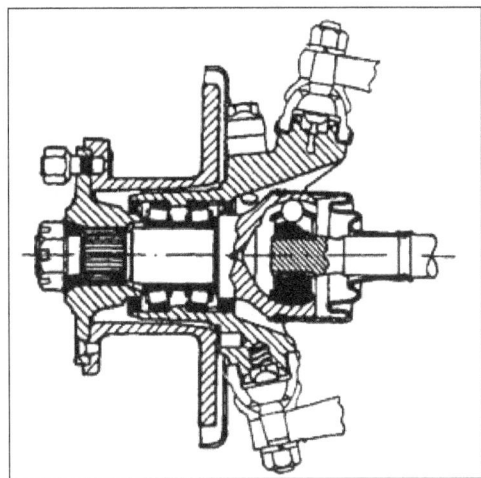

Front hub with front-wheel drive.

Bearing Loads on the Front Axle

Figure illustrates the forces and the reactions on steering knuckle when the vehicle is at rest. The thrust load and the knuckle-pin-bearing load can be expressed in terms of the reaction of wheel on wheel spindle. Let, Rw = The reaction of the wheel on the spindle acting vertically through the centre of contact of tyre on ground. Rt = The load on the thrust bearing Ru = The load on the upper knucklepin bearing Rl = The load on lower knucklepin 'B' and 'C represent the centres of lower and upper knuckle-pin bearings respectively. 'A' is a point on the spindle axis in the centre plane of the wheel.

Then $\sum M_c = 0$ gives, $R_{wc} - R_l(d+e) = 0$

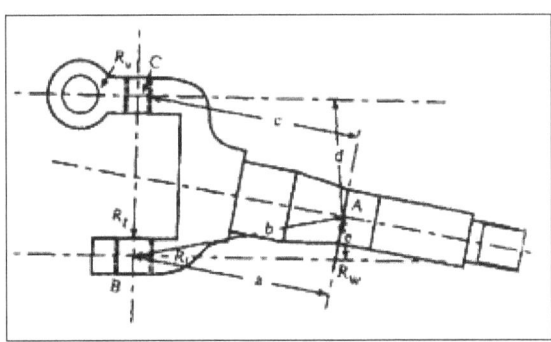

Forces and reaction on steering knuckle.

or, $R_l = \dfrac{c}{d+e} R_w$

Similarly, $\sum M_B = 0$ gives, $R_w a - R_u(d+e) = 0$

or $R_u = \dfrac{a}{a+e} R_w$

and, $\sum M_A = 0$ gives, $R_t b - R_{l_e} - R_u d = 0$

or $R_t b = R_{l_e} + R_u\, d$

Substituting the expression of R_l and R_u,

$$R_t = \frac{ce+ad}{b(d+e)} R_w.$$

The other loads acting on knuckle-pin bearing are those due to the rolling resistance and road shocks. These loads are proportional to the static load and hence can be accounted for.

Bearing Lubrication

The whole hub assembly is partially packed with grease. A radial-lip grease-seal is pressed inside the hub next to the larger bearing. The lip of the seal faces this bearing and fits over the cylindrically machined surface on the stub-axle. Over packing bearing hubs with grease causes churning and develops very high running temperatures. Due to high temperatures and excessive mechanical working the grease breaks down and becomes soft. The centrifugal force acting on the grease in the hub pushes soft grease outwards, so that it flows towards the lip seal and seeps past the seal lips. Bearing greases are based on mineral oil and to thicken the oil soaps of calcium, sodium, or lithium compounds are added. The consistency of a grease depends on the viscosity of the base oil as well as on the structure and properties of the metallic soap added. The three most important properties of a grease includes (i) the melting temperature, at which the grease loses its semi-solid state, (ii) its resistance to contamination and dilution by water, and (Hi) its ability to withstand mechanical working before the grease breaks. The relative operating properties of commonly used bearing greases are provided in Table.

Table: Operating properties of commonly used greases.

Types of grease soap	Upper temperature limits		Water resistance	Mechanical stability
	Short duration K	Continuous K		
Calcium	353	323	Good	Fair
Sodium	393	353	Poor	Good
Lithium	423	393	Good	Good

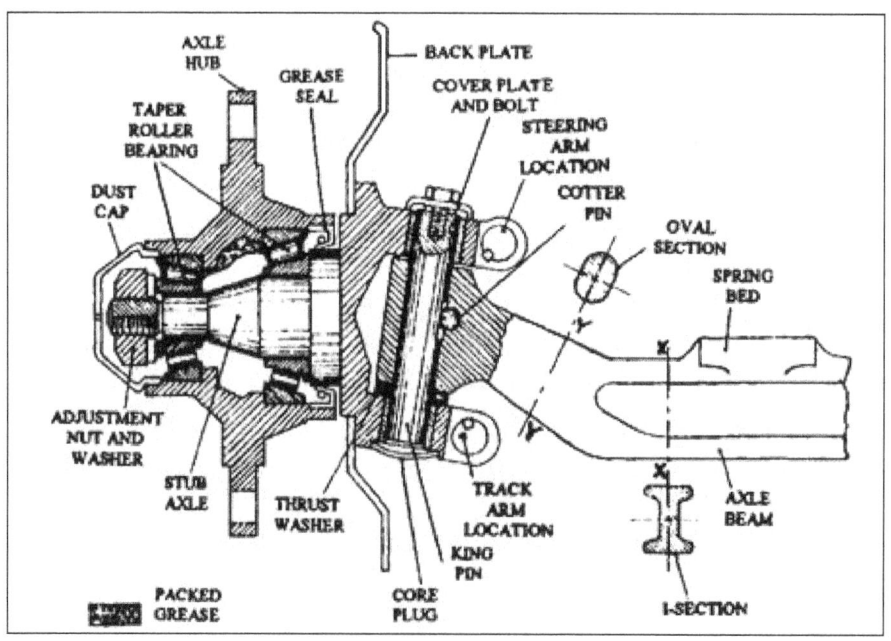

Axle-beam and stub-axle assembly.

For checking the bearing and seal, the split-pin and the castellated nut on the end of the stub-axle are removed and the hub assembly is withdrawn. All parts are thoroughly washed in paraffin, and the bearing rollers and tracks are examined for wear and damage. Before replacing the hub, the inside of the hub and the bearings are applied with grease.

Steering System

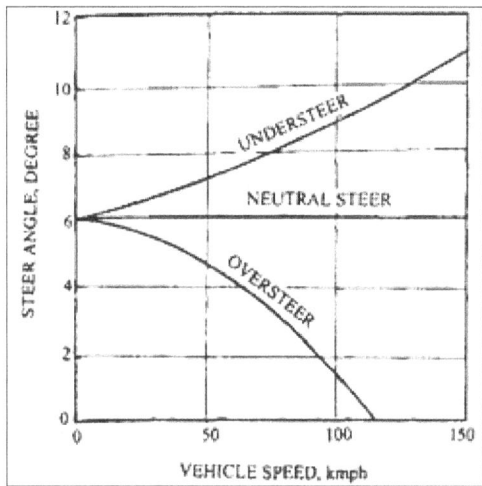

Relationship of steer angle speed and vehicle speed for various steering conditions.

The function of a steering system is to convert the rotary movement of the steering wheel in driver's hand into the angular turn of the front wheels on road. Additionally, the steering system should provide mechanical advantage over front wheel steering knuckles, offering driver an easy turning of front wheels with minimum effort in any desired direction. The main causes of stiff steering include (a) insufficient lubrication of the king-pins or steering linkage, (b) tyre pressure too low, (c) wheels out of track, i.e. toe-in not correct, and (d) stiffness in the steering column itself, caused by lack of lubricant or over tightening.

The steering system is designed to enable the driver to control and continuously adjust the steered path of the vehicle. Also it provides a positive response to whatever direction the driver may makes on the steering wheel. To achieve these objectives, a suitable mechanical linkage is incorporated between the front steered road-wheels and the driver's steering-wheel.

This mechanism operates effectively under all normal conditions without interfering with the wheel traction or with the suspension movement.

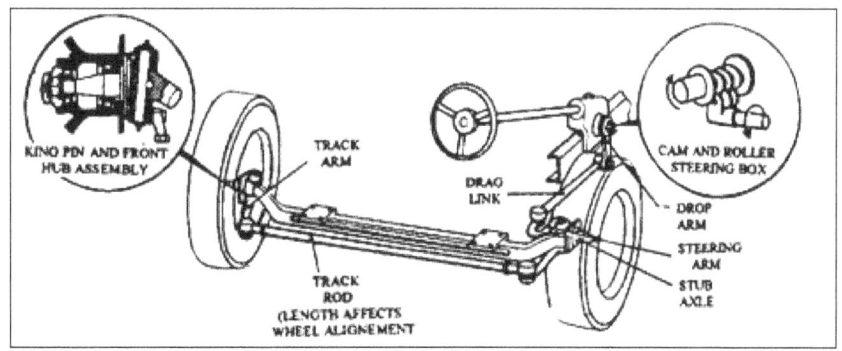

Steering layout of light truck.

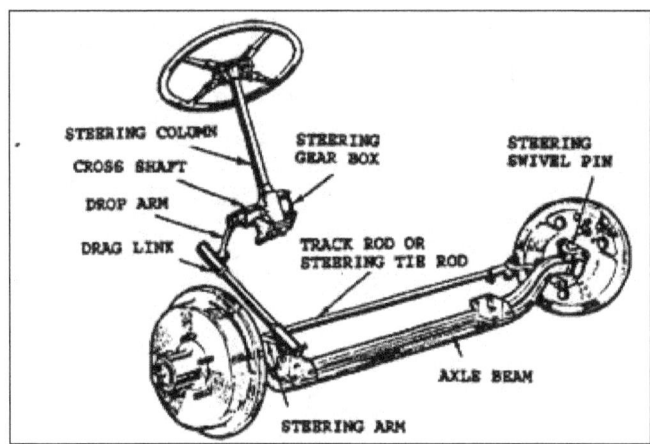

Schematice view of steering linkage.

The steering linkage shown in Fig. (schematic view) performs the above functions. When driver turns the steering wheel, motion is transmitted down through the steering tube to the steering gear. The steering tube revolves inside the steering column. The steering gear changes the direction of motion and increases the turning force applied by driver at the steering wheel in accordance with the gear ratio. The gear rotates the steering arm (Pitman arm), which transfers the motion to the steering knuckles through the steering gear connecting rod, tie-rod, and knuckle arms. This type of linkage is called the relay steering linkage. The layout of any steering linkage depends largely on the type of vehicle to which it is fitted. A commercial vehicle uses a rigid axle beam front suspension steering system. A car generally relies on independent front suspension steering system.

Steering layout for car.

Axle-beam Suspension Steering System

This steering system incorporates a steering-wheel to impart motion to the steering-box which transfers the steering effort through the drop-arm and drag-link directly to one of the two stub-axles pivoting at the ends of the axle-beam. Both the stub-axles are joined together by a track-rod. Figure illustrates the axle beam steering layout in one of its views and the functions of the components are as follows :
Steering Box. The steering box uses a reduction gear which provides a much larger force to the steering linkage with only a small effort. Simultaneously, the degree of stub-axle movement is decreased for a given angular movement of the steering wheel so that the oversensitivity of the steering with respect to driver's touch on the wheel is reduced.

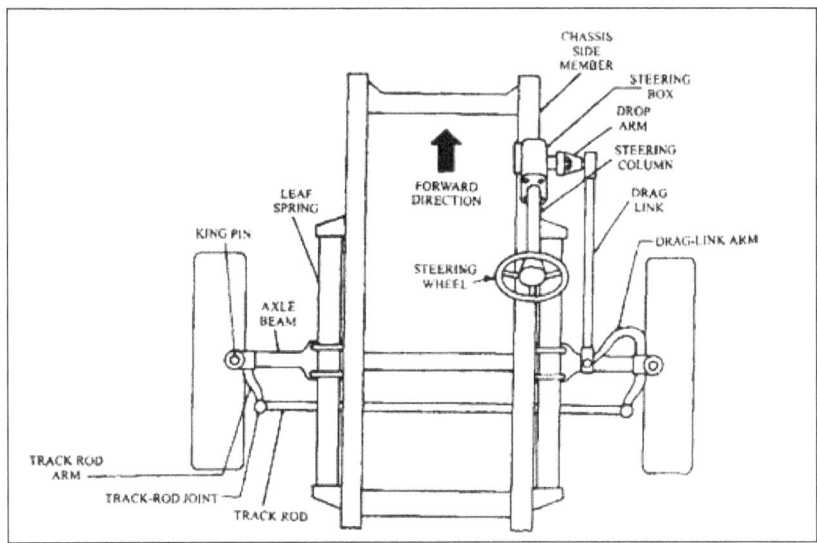

Axle beam steering linkage with longitudinal located drag link.

Drop-arm

This forged lever-arm is bolted on to a tapered steering-box output rocker-shaft and it hangs or drops downwards. It imparts a circular-arc movement to the drag-link through its swing action.

Drag-link

This tubular rod converts the circular movement of the drop-arm into a linear push or pull motion of the drag-link arm, attached rigidly to one of the stub-axles. A ball-joint is fitted at each end of the rod so that a relative movement is provided in planes. Figure shows an alternative transverse drag link layout suitable for cross country applications.

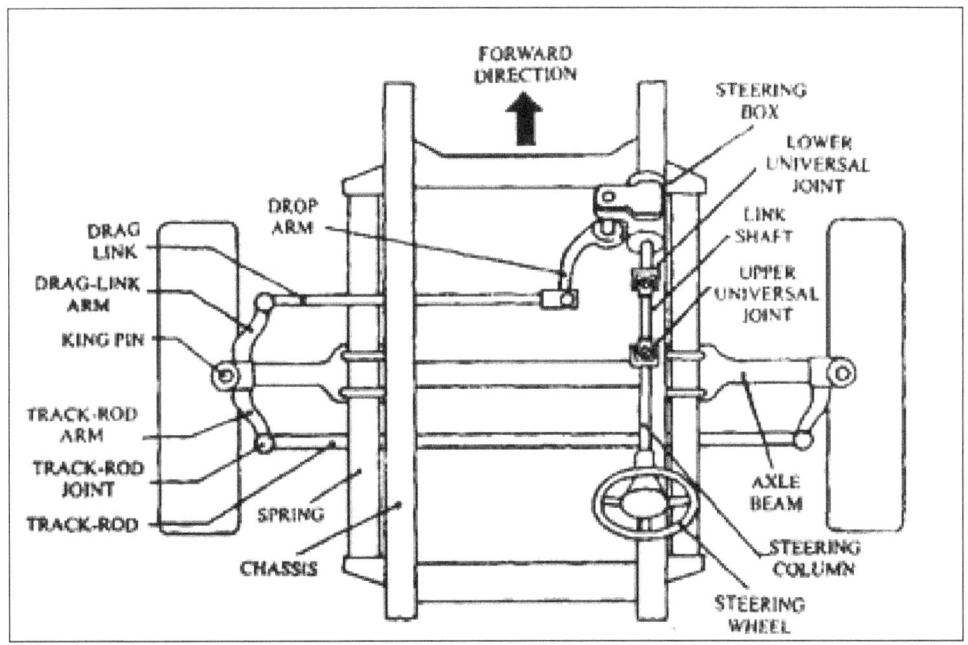

Axle-beam steering linkage with transverse located drag link.

Drag-link Arm

This arm joints the drag-link to one of the stub-axles and provides sufficient leverage to convert the linear movement of the drag-link to an angular movement about the stub-axle king-pin.

Drag-link Suspension and Steering Interaction

The axle-beam pivots about the fixed front shackle-pins and moves up and down in a circular arc. Also the drag-link pivots about the drop-arm ball-joint during any vertical movement of the axle. When the effective arc radius of the axle movement and the drag-link arm end are approximately equal, the suspension axle movement relative to the chassis is independent of the vehicle's steered path. If a slight difference exists during deflection of the suspension, then the drag-link proportionally increases or decreases the relative angular position of the stub-axle about the king-pin. This causes the steering to continuously twitch or jerk when encountering rough surface conditions.

Stub-axles

The stub-axle is a short axle-shaft to which one steered road-wheel is mounted. It uses two extended horizontal prongs that fit over the ends of the axle-beam. The king-pin, a short circular bar, passes vertically through both prongs and the eye of the axle-beam to form the hinge pivot. The stub-axle acts as the wheel axle as well as the pivot support member in the horizontal plane.

Track-rod Arms

Each stub-axle uses a forged track-rod arm bolted approximately at right angles to the wheel axis in the horizontal plane. This arm provides the leverage to rotate the stub-axle about the king-pin. This rotary movement is transferred to the other stub-axle through the track-rod.

Track-rod

A tubular track-rod spans the wheel track and pivots together the two stub-axles. The ends of this rod carry ball-joints, which in turn are bolted to the track-rod arms of each stub axle. These ball-joints are allowed to move only in the horizontal plane. The drag-link movement is either a pull or a push action and rotates one of the stub-axles. This motion is transferred to the other stub-axle through the track-rod.

Independent Suspension Steering System

In the rigid-beam suspension, the stub-axle is pivoted at each end of the axle-beam. Consequently the relative movement is permitted only in the horizontal plane due to which effective track-rod length is not affected by any vertical suspension deflection. Independent-suspension steering, on the other hand, copes with up a down movement of each stub-axle independent of the other due to which the distance between track-rod-arm ball-joint centres varies continually. Therefore, if a single track-rod joins the two stub-axles together, the slightest bump or rebound tends to pull both stub-axle arms at once and thus interferes with the steering-track toe-in or toe-out. To overcome the problem of the changing distance between track-rod-arm ball-joint centres, a three-piece track-rod is used. The centre portion of the track rod may be a relay-rod suspended between the steering-box drop-arm and an idler arm fixed to the body structure. Also the centre portion may from the track shaft of a rack-and-pinion steering-box. In both the cases, this part moves only in the horizontal plane. Movement in the vertical plane is provided by the two outer connecting rods, known as tie-rods. The tie-rods swing about the ball-joints placed at the end of the middle track rod member. In earlier designs, independent suspension steering incorporated stub-axles and king-pin pivots similar to those used with the axle-beam. But current systems use ball-swivel joints for the stub axle pivot and are also spaced further apart. Large cars normally use the system shown in Figure. When the steering-wheel is acted, the drop-arm conveys movement to the relav-rod, which inturn transmits this motion to both tie-rods and stub-axles. The drop-arm and idler-arm relay joints provide movement only in the horizontal plane. The tie-rod joints provide movement in both the horizontal and vertical planes.

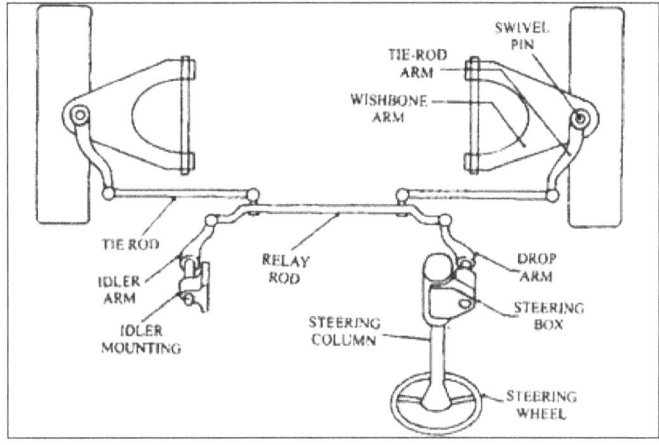

Split track-rod with relay-rod and idler steering linkage layout.

The most popular steering system used for small and medium cars is shown in Fig. This type of steering box has a rack-and-pinion housing bolted along the body cross-member. The angular movement of steering wheel is converted to a linear to-and-fro movement of the rack. Each end of the rack shaft is attached to a tie-rod by means of a ball-and-socket joint. The outer tie-rod ends also use ball-joints, which are bolted to the stub-axle track-rod arms. The rack shaft thus provides the transverse steering thrust and the tie-rod ball joints allow pivoting in two planes.

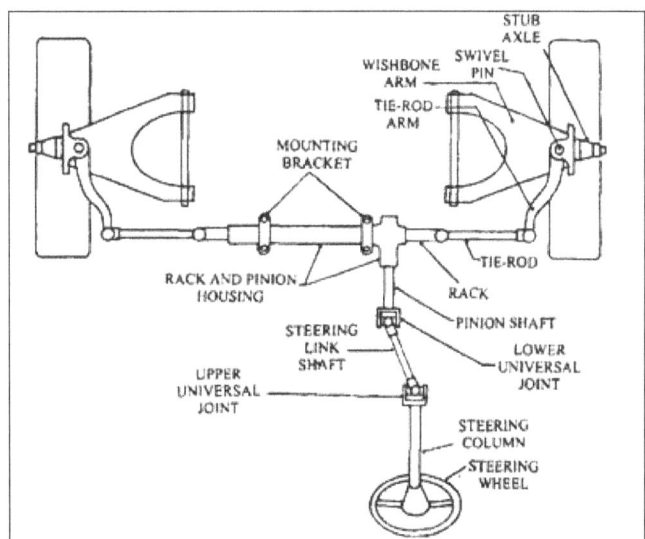

Rack-and-pinion steering linkage layout.

Suspension System

Suspension is the system of tires, tire air, springs, shock absorbers and linkages that connects a vehicle to its wheels and allows relative motion between the two. Suspension systems must support both road holding/handling and ride quality, which are at odds with each other. The tuning of suspensions involves finding the right compromise. It is important for the suspension to keep the road wheel in contact with the road surface as much as possible, because all the road or ground forces acting on the vehicle do so through the contact patches of the tires. The suspension also protects the vehicle itself and any cargo or luggage from damage and wear. The design of front and rear suspension of a car may be different.

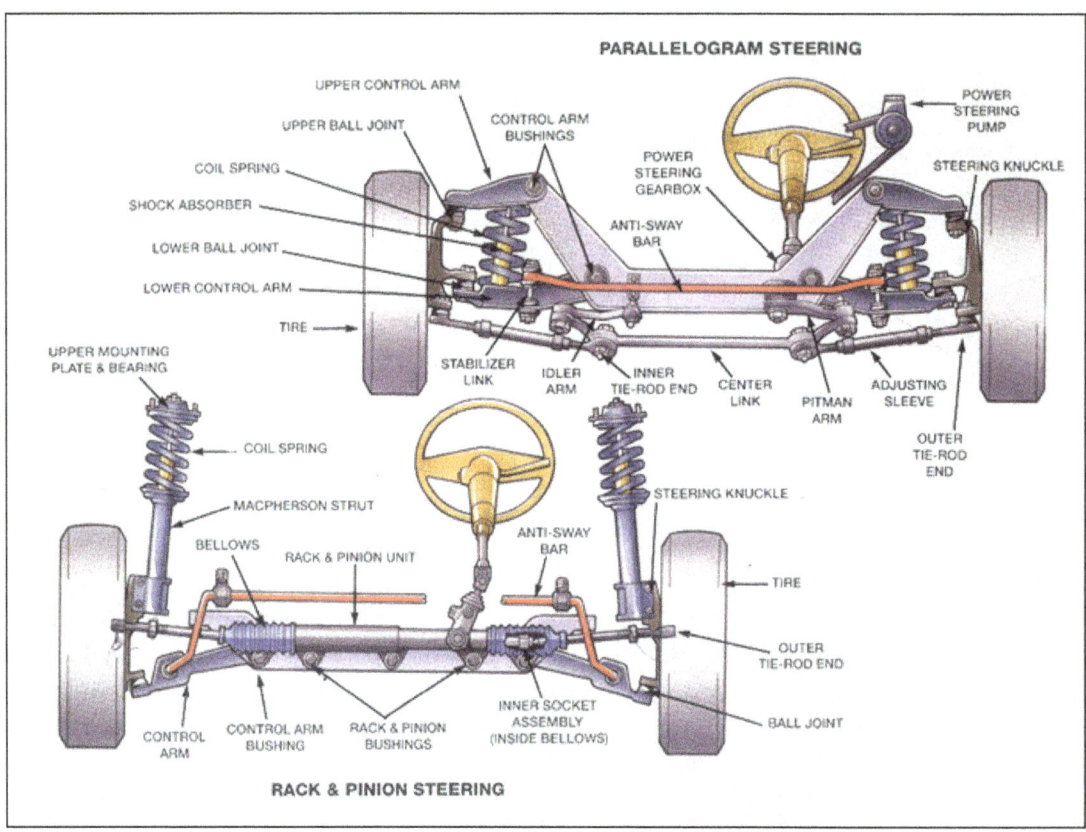

Difference between Rear Suspension and Front Suspension

Any four wheel vehicle needs suspension for both the front wheels and the rear suspension, but in two wheel drive vehicles there can be a very different configuration. For front-wheel drive cars, rear suspension has few constraints and a variety of beam axles and independent suspensions are used. For rear-wheel drive cars, rear suspension has many constraints and the development of the superior but more expensive independent suspension layout has been difficult. Four-wheel drive often has suspensions that are similar for both the front and rear wheels.

Henry Ford's Model T used a torque tube to restrain this force, for his differential was attached to the chassis by a lateral leaf spring and two narrow rods. The torque tube surrounded the true driveshaft and exerted the force to its ball joint at the extreme rear of the transmission, which was attached to the engine. A similar method was used in the late 1930s by Buick and by Hudson's bathtub car in 1948, which used helical springs which could not take fore-and-aft thrust.

The Hotchkiss drive, invented by Albert Hotchkiss, was the most popular rear suspension system used in American cars from the 1930s to the 1970s. The system uses longitudinal leaf springs attached both forward and behind the differential of the live axle. These springs transmit the torque to the frame. Although scorned by many European car makers of the time, it was accepted by American car makers because it was inexpensive to manufacture. Also, the dynamic defects of this design were suppressed by the enormous weight of US passenger vehicles before implementation of the Corporate Average Fuel Economy standard.

Another Frenchman invented the De Dion tube, which is sometimes called "semi-independent". Like a true independent rear suspension, this employs two universal joints or their equivalent from the centre of the differential to each wheel. But the wheels cannot entirely rise and fall independently of each other; they are tied by a yoke that goes around the differential, below and behind it. This method has had little use in the United States. Its use around 1900 was probably due to the poor quality of tires, which wore out quickly. By removing a good deal of unsprung weight, as independent rear suspensions do, it made them last longer.

Rear wheel drive vehicles today frequently use a fairly complex fully independent, multi-link suspension to locate the rear wheels securely while providing decent ride quality.

Spring, Wheel and Roll Rates

Spring Rate

The spring rate (or suspension rate) is a component in setting the vehicle's ride height or its location in the suspension stroke. When a spring is compressed or stretched, the force it exerts is proportional to its change in length. The spring rate or spring constant of a spring is the change in the force it exerts, divided by the change in deflection of the spring. Vehicles which carry heavy loads will often have heavier springs to compensate for the additional weight that would otherwise collapse a vehicle to the bottom of its travel (stroke). Heavier springs are also used in performance applications where the loading conditions experienced are more extreme.

Springs that are too hard or too soft cause the suspension to become ineffective because they fail to properly isolate the vehicle from the road. Vehicles that commonly experience suspension loads heavier than normal have heavy or hard springs with a spring rate close to the upper limit for that vehicle's weight. This allows the vehicle to perform properly under a heavy load when control is limited by the inertia of the load. Riding in an empty truck used for carrying loads can be uncomfortable for passengers because of its high spring rate relative to the weight of the vehicle. A race car would also be described as having heavy springs and would also be uncomfortably bumpy. However, even though we say they both have heavy springs, the actual spring rates for a 2,000 lb (910 kg) race car and a 10,000 lb (4,500 kg) truck are very different. A luxury car, taxi, or passenger bus would be described as having soft springs. Vehicles with worn out or damaged springs ride lower to the ground which reduces the overall amount of compression available to the suspension and increases the amount of body lean. Performance vehicles can sometimes have spring rate requirements other than vehicle weight and load.

Mathematics of the Spring Rate

Spring rate is a ratio used to measure how resistant a spring is to being compressed or expanded during the spring's deflection. The magnitude of the spring force increases as deflection increases according to Hooke's Law. Briefly, this can be stated as:

$$F = -kx$$

where:

> F is the force the spring exerts
>
> k is the spring rate of the spring
>
> x is the deflection of the spring from its equilibrium position (i.e., when no force is applied on the spring)

The negative sign indicates the direction of applied force and force exerted by spring are opposite. Spring rate is confined to a narrow interval by the weight of the vehicle, load the vehicle will carry, and to a lesser extent by suspension geometry and performance desires.

Spring rates typically have units of N/mm (or lbf/in). An example of a linear spring rate is 500 lbf/in. For every inch the spring is compressed, it exerts 500 lbf. A non-linear spring rate is one for which the relation between the spring's compression and the force exerted cannot be fitted adequately to a linear model. For example, the first inch exerts 500 lbf force, the second inch exerts an additional 550 lbf (for a total of 1050 lbf), the third inch exerts another 600 lbf (for a total of 1650 lbf). In contrast a 500 lbf/in linear spring compressed to 3 inches will only exert 1500 lbf.

The spring rate of a coil spring may be calculated by a simple algebraic equation or it may be measured in a spring testing machine. The spring constant k can be calculated as follows:

$$k = \frac{d^4 G}{8ND^3}$$

where d is the wire diameter, G is the spring's shear modulus (e.g., about 12,000,000 lbf/in² or 80 GPa for steel), N is the number of wraps and D is the diameter of the coil.

Wheel Rate

Wheel rate is the effective spring rate when measured at the wheel as opposed to simply measuring the spring rate alone.

Wheel rate is usually equal to or considerably less than the spring rate. Commonly, springs are mounted on control arms, swing arms or some other pivoting suspension member. Consider the example above where the spring rate was calculated to be 500 lbs/inch (87.5 N/mm), if you were to move the wheel 1 in (2.5 cm) (without moving the car), the spring more than likely compresses a smaller amount. Let's assume the spring moved 0.75 in (19 mm), the lever arm ratio would be 0.75:1. The wheel rate is calculated by taking the square of the ratio (0.5625) times the spring rate, thus obtaining 281.25 lbs/inch (49.25 N/mm). Squaring the ratio is because the ratio has two effects on the wheel rate. The ratio applies to both the force and distance traveled.

Wheel rate on independent suspension is fairly straightforward. However, special consideration must be taken with some non-independent suspension designs. Take the case of the straight axle. When viewed from the front or rear, the wheel rate can be measured by the means above. Yet, because the wheels are not independent, when viewed from the side under acceleration or braking, the pivot point is at infinity (because both wheels have moved) and the spring is directly inline with the wheel contact patch. The result is often that the effective wheel rate under cornering is different from what it is under acceleration and braking. This variation in wheel rate may be minimised by locating the spring as close to the wheel as possible.

Wheel rates are usually summed and compared with the sprung mass of a vehicle to create a "ride rate" and corresponding suspension natural frequency in ride (also referred to as "heave"). This can be useful in creating a metric for suspension stiffness and travel requirements for a vehicle.

Roll Rate

Roll rate is analogous to a vehicle's ride rate, but for actions that include lateral accelerations, causing a vehicle's sprung mass to roll about its roll axis. It is expressed as torque per degree of roll of the vehicle sprung mass. It is influenced by factors including but not limited to vehicle sprung mass, track width, CG height, spring and damper rates, roll centre heights of front and rear, anti-roll bar stiffness and tire pressure/construction. The roll rate of a vehicle can, and usually does, differ front to rear, which allows for the tuning ability of a vehicle for transient and steady state handling. The roll rate of a vehicle does not change the total amount of weight transfer on the vehicle, but shifts the speed and percentage of weight transferred on a particular axle to another axle through the vehicle chassis. Generally, the higher the roll rate on an axle of a vehicle, the faster and higher percentage the weight transfer on that axle.

Roll Couple Percentage

Roll couple percentage is a simplified method of describing lateral load transfer distribution front to rear, and subsequently handling balance. It is the effective wheel rate, in roll, of each axle of the vehicle as a ratio of the vehicle's total roll rate. It is commonly adjusted through the use of anti-roll bars, but can also be changed through the use of different springs.

Weight Transfer

Weight transfer during cornering, acceleration or braking is usually calculated per individual wheel and compared with the static weights for the same wheels.

The total amount of weight transfer is only affected by four factors: the distance between wheel centers (wheelbase in the case of braking, or track width in the case of cornering) the height of the center of gravity, the mass of the vehicle, and the amount of acceleration experienced.

The speed at which weight transfer occurs as well as through which components it transfers is complex and is determined by many factors including but not limited to roll center height, spring and damper rates, anti-roll bar stiffness and the kinematic design of the suspension links. In most conventional applications, when weight is transferred through intentionally compliant elements such as springs, dampers and anti-roll bars, the weight transfer is said to be "elastic", while the weight which is transferred through more rigid suspension links such as A-arms and toe links is said to be "geometric".

Unsprung Weight Transfer

Unsprung weight transfer is calculated based on the weight of the vehicle's components that are not supported by the springs. This includes tires, wheels, brakes, spindles, half the control arm's weight and other components. These components are then (for calculation purposes) assumed to be connected to a vehicle with zero sprung weight. They are then put through the same dynamic loads. The weight transfer for cornering in the front would be equal to the total unsprung front weight times the G-Force times the front unsprung center of gravity height divided by the front track width. The same is true for the rear.

Sprung Weight Transfer

Sprung weight transfer is the weight transferred by only the weight of the vehicle resting on the springs, not the total vehicle weight. Calculating this requires knowing the vehicle's sprung weight (total weight less the unsprung weight), the front and rear roll center heights and the sprung center of gravity height (used to calculate the roll moment arm length). Calculating the front and rear sprung weight transfer will also require knowing the roll couple percentage.

The roll axis is the line through the front and rear roll centers that the vehicle rolls around during cornering. The distance from this axis to the sprung center of gravity height is the roll moment arm length. The total sprung weight transfer is equal to the G-force times the sprung weight times the roll moment arm length divided by the effective track width. The front sprung weight transfer is calculated by multiplying the roll couple percentage times the total sprung weight transfer. The rear is the total minus the front transfer.

Jacking Forces

Jacking forces are the sum of the vertical force components experienced by the suspension links. The resultant force acts to lift the sprung mass if the roll center is above ground, or compress it if underground. Generally, the higher the roll center, the more jacking force is experienced.

Other Properties

Travel

Travel is the measure of distance from the bottom of the suspension stroke (such as when the vehicle is on a jack and the wheel hangs freely) to the top of the suspension stroke (such as when the vehicle's wheel can no longer travel in an upward direction toward the vehicle). Bottoming or lifting a wheel can cause serious control problems or directly cause damage. "Bottoming" can be caused by the suspension, tires, fenders, etc. running out of space to move or the body or other components of the car hitting the road. The control problems caused by lifting a wheel are less severe if the wheel lifts when the spring reaches its unloaded shape than they are if travel is limited by contact of suspension members. Many off-road vehicles, such as desert racers, use straps called "limiting straps" to limit the suspensions downward travel to a point within safe limits for the linkages and shock absorbers. This is necessary, since these trucks are intended to travel over very rough terrain at high speeds, and even become airborne at times. Without something to limit the travel, the suspension bushings would take all the force when the suspension reaches "full droop", and it can even cause the coil springs to come out of their "buckets" if they are held in by compression forces only. A limiting strap is a simple strap, often nylon of a predetermined length, that stops the downward movement at a preset point before the theoretical maximum travel is reached. The opposite of this is the "bump-stop", which protects the suspension and vehicle (as well as the occupants) from violent "bottoming" of the suspension, caused when an obstruction (or hard landing) causes the suspension to run out of upward travel without fully absorbing the energy of the stroke. Without bump-stops, a vehicle that "bottoms out" will expe-rience a very hard shock when the suspension contacts the bottom of the frame or body, which is transferred to the occupants and every connector and weld on the vehicle. Factory vehicles often come with plain rubber "nubs" to absorb the worst of the forces, and insulate the shock. A desert race vehicle, which must routinely absorb far higher impact forces, may be provided with pneumatic or hydropneumatic bump-stops. These are essentially miniature shock absorbers (dampeners) that are fixed to the vehicle in a location such that the suspension will contact the end of the piston when it nears the upward travel limit. These absorb the impact far more effectively than a solid rubber bump-stop will, essential because a rubber bump-stop is considered a "last-ditch" emergency insulator for the occasional accidental bottoming of the suspension; it is entirely insufficient to absorb repeated and heavy bottomings such as a high-speed off-road vehicle encounters.

Damping

Damping is the control of motion or oscillation, as seen with the use of hydraulic gates and valves in a vehicle's shock absorber. This may also vary, intentionally or unintentionally. Like spring rate, the optimal damping for comfort may be less than for control.

Damping controls the travel speed and resistance of the vehicle's suspension. An undamped car will oscillate up and down. With proper damping levels, the car will settle back to a normal state in a minimal amount of time. Most damping in modern vehicles can be controlled by increasing or decreasing the resistance to fluid flow in the shock absorber.

Camber Control

Camber changes due to wheel travel, body roll and suspension system deflection or compliance. In general, a tire wears and brakes best at -1 to -2° of camber from vertical. Depending on the tire and the road surface, it may hold the road best at a slightly different angle. Small changes in camber, front and rear, can be used to tune handling. Some race cars are tuned with -2 to -7° camber depending on the type of handling desired and the tire construction. Often, too much camber will result in the decrease of braking performance due to a reduced contact patch size through excessive camber variation in the suspension geometry. The amount of camber change in bump is determined by the instantaneous front view swing arm (FVSA) length of the suspension geometry, or in other words, the tendency of the tire to camber inward when compressed in bump.

Roll Center Height

Roll center height is a product of suspension instant center heights and is a useful metric in analyzing weight transfer effects, body roll and front to rear roll stiffness distribution. Conventionally, roll stiffness distribution is tuned adjusting antiroll bars rather than roll center height (as both tend to have a similar effect on the sprung mass), but the height of the roll center is significant when considering the amount of jacking forces experienced.

Instant Center

Due to the fact that the wheel and tire's motion is constrained by the suspension links on the vehicle, the motion of the wheel package in the front view will scribe an imaginary arc in space with an "instantaneous center" of rotation at any given point along its path. The instant center for any wheel package can be found by following imaginary lines drawn through the suspension links to their intersection point.

A component of the tire's force vector points from the contact patch of the tire through instant center. The larger this component is, the less suspension motion will occur. Theoretically, if the resultant of the vertical load on the tire and the lateral force generated by it points directly into the instant center, the suspension links will not move. In this case, all weight transfer at that end of the vehicle will be geometric in nature. This is key information used in finding the force-based roll center as well.

In this respect the instant centers are more important to the handling of the vehicle than the kinematic roll center alone, in that the ratio of geometric to elastic weight transfer is determined by the forces at the tires and their directions in relation to the position of their respective instant centers.

Anti-dive and Anti-squat

Anti-dive and anti-squat are percentages that indicate the degree to which the front dives under braking and the rear squats under acceleration. They can be thought of as the counterparts for braking and acceleration, as jacking forces are to cornering. The main reason for the difference is due to the different design goals between front and rear suspension, whereas suspension is usually symmetrical between the left and right of the vehicle.

The method of determining the anti-dive or anti-squat depends on whether the suspension linkages react to the torque of braking and accelerating. For example, with inboard brakes and half-shaft driven rear wheels, the suspension linkages do not, but with outboard brakes and a swing-axle driveline, they do.

To determine the percentage of front suspension braking anti-dive for outboard brakes, it is first necessary to determine the tangent of the angle between a line drawn, in side view, through the front tire patch and the front suspension instant center, and the horizontal. In addition, the percentage of braking effort at the front wheels must be known. Then, multiply the tangent by the front wheel braking effort percentage and divide by the ratio of the center of gravity height to the wheelbase. A value of 50% would mean that half of the weight transfer to the front wheels, during braking, is being transmitted through the front suspension linkage and half is being transmitted through the front suspension springs.

For inboard brakes, the same procedure is followed but using the wheel center instead of contact patch center.

Forward acceleration anti-squat is calculated in a similar manner and with the same relationship between percentage and weight transfer. Anti-squat values of 100% and more are commonly used in drag racing, but values of 50% or less are more common in cars that have to undergo severe braking. Higher values of anti-squat commonly cause wheel hop during braking. It is important to note that, while the value of 100% means that all of the weight transfer is being carried through the suspension linkage. However, this does not mean that the suspension is incapable of carrying additional loads (aerodynamic, cornering, etc.) during an episode of braking or forward acceleration. In other words, no "binding" of the suspension is to be implied.

Flexibility and Vibration Modes of the Suspension Elements

In some modern cars, the flexibility is mainly in the rubber bushings, which are subject to decay over time. For high-stress suspensions, such as off-road vehicles, polyurethane bushings are available, which offer more longevity under greater stresses. However, due to weight and cost considerations, structures are not made more rigid than necessary. Some vehicles exhibit detrimental vibrations involving the flexing of structural parts, such as when accelerating while turning sharply. Flexibility of structures such as frames and suspension links can also contribute to springing, especially to damping out high frequency vibrations. The flexibility of wire wheels contributed to their popularity in times when cars had less advanced suspensions.

Load Levelling

Automobiles can be heavily laden with luggage, passengers, and trailers. This loading will cause a vehicle's tail to sink downwards. Maintaining a steady chassis level is essential to achieving the proper handling the vehicle was designed for. Oncoming drivers can be blinded by the headlight beam. Self-levelling suspension counteracts this by inflating cylinders in the suspension to lift the chassis higher.

Isolation from High Frequency Shock

For most purposes, the weight of the suspension components is unimportant, but at high frequencies, caused by road surface roughness, the parts isolated by rubber bushings act as a multistage filter to suppress noise and vibration better than can be done with only the tires and springs. (The springs work mainly in the vertical direction.)

Contribution to Unsprung Weight and Total Weight

These are usually small, except that the suspension is related to whether the brakes and differential(s) are sprung.

This is the main functional advantage of aluminum wheels over steel wheels. Aluminum suspension parts have been used in production cars, and carbon fiber suspension parts are common in racing cars.

Space Occupied

Designs differ as to how much space they take up and where it is located. It is generally accepted that MacPherson struts are the most compact arrangement for front-engined vehicles, where space between the wheels is required to place the engine.

Inboard brakes (which reduce unsprung weight) are probably avoided more due to space considerations than to cost.

Force Distribution

The suspension attachment must match the frame design in geometry, strength and rigidity.

Air Resistance

Certain modern vehicles have height adjustable suspension in order to improve aerodynamics and fuel efficiency. Modern formula cars that have exposed wheels and suspension typically use streamlined tubing rather than simple round tubing for their suspension arms to reduce aerodynamic drag. Also typical is the use of rocker arm, push rod, or pull rod type suspensions that, among other things, place the spring/damper unit inboard and out of the air stream to further reduce air resistance.

Cost

Production methods improve, but cost is always a factor. The continued use of the solid rear axle, with unsprung differential, especially on heavy vehicles, seems to be the most obvious example.

Springs and Dampers

Most conventional suspensions use passive springs to absorb impacts and dampers (or shock absorbers) to control spring motions.

Some notable exceptions are the hydropneumatic systems, which can be treated as an integrated unit of gas spring and damping components, used by the French manufacturer Citroën and the hydrolastic, hydragas and rubber cone systems used by the British Motor Corporation, most notably on the Mini. A number of different types of each have been used:

Passive Suspensions

Traditional springs and dampers are referred to as passive suspensions — most vehicles are suspended in this manner.

Springs

The majority of land vehicles are suspended by steel springs, of these types:

- Leaf spring – AKA Hotchkiss, Cart, or semi-elliptical spring.
- Torsion bar suspension.
- Coil spring.

Automakers are aware of the inherent limitations of steel springs, that they tend to produce undesirable oscillations, and have developed other types of suspension materials and mechanisms in attempts to improve performance:

- Rubber bushing.

- Gas under pressure - air spring.

- Gas and hydraulic fluid under pressure - hydropneumatic suspension and oleo strut.

Dampers or Shock Absorbers

The shock absorbers damp out the (otherwise simple harmonic) motions of a vehicle up and down on its springs. They also must damp out much of the wheel bounce when the unsprung weight of a wheel, hub, axle and sometimes brakes and differential bounces up and down on the springiness of a tire.

Semi-active and Active Suspensions

If the suspension is externally controlled then it is a semi-active or active suspension — the suspension is reacting to signals from an electronic controller.

For example, a hydropneumatic Citroën will "know" how far off the ground the car is supposed to be and constantly reset to achieve that level, regardless of load. It will not instantly compensate for body roll due to cornering however. Citroën's system adds about 1% to the cost of the car versus passive steel springs.

Semi-active suspensions include devices such as air springs and switchable shock absorbers, various self-levelling solutions, as well as systems like hydropneumatic, hydrolastic, and hydragas suspensions. Toyota introduced switchable shock absorbers in the 1983 Soarer. Delphi currently sells shock absorbers filled with a magneto-rheological fluid, whose viscosity can be changed electromagnetically, thereby giving variable control without switching valves, which is faster and thus more effective.

Fully active suspension systems use electronic monitoring of vehicle conditions, coupled with the means to change the behavior of the vehicle suspensionin real time to directly control the motion of the car. Lotus Cars developed several prototypes, from 1982 onwards, and introduced them to Formula One, where they have been fairly effective, but have now been banned. Nissan introduced a low bandwidth active suspension in circa 1990 as an option that added an extra 20% to the price of luxury models. Citroën has also developed several active suspension models. A recently publicised fully active system from Bose Corporation uses linear electric motors (i.e., solenoids) in place of hydraulic or pneumatic actuators that have generally been used up until recently. Mercedes introduced an active suspension system called Active Body Control in its top-of-the-line Mercedes-Benz CL-Class in 1999.

Several electromagnetic suspensions have also been developed for vehicles. Examples include the electromagnetic suspension of Bose, and the electromagnetic suspension developed by prof. Laurentiu Encica. In addition, the new Michelin wheel with embedded suspension working on an electromotor is also similar.

With the help of control system, various semi-active/active suspensions realize an improved design compromise among different vibrations modes of the vehicle, namely bounce, roll, pitch and warp modes. However, the applications of these advanced suspensions are constrained by the cost, packaging, weight, reliability, and other challenges.

Interconnected Suspensions

Interconnected suspension, unlike semi-active/active suspensions, could easily decouple different vehicle vibration modes in a passive manner. The interconnections can be realized by various means, such as mechanical, hydraulic and pneumatic. Anti-roll bars are one of the typical examples of mechanical interconnections, while it has been stated that fluidic interconnections offer greater potential and flexibility in improving both the stiffness and damping properties.

Considering the considerable commercial potentials of hydro-pneumatic technology (Corolla, 1996), interconnected hydropneumatic suspensions have also been explored in some recent studies, and their potential benefits in enhancing vehicle ride and handling have been demonstrated. The control system can also be used for further improving performance of interconnected suspensions. Apart from academic research, an Australian company, Kinetic, had some success (WRC: 3 Championships, Dakar Rally: 2 Championships, Lexus GX470 2004 4×4 of the year with KDSS, 2005 PACE award) with various passive or semi-active systems, which generally decouple at least two vehicle modes (roll, warp (articulation), pitch and heave (bounce)) to simultaneously control each mode's stiffness and damping, by using interconnected shock absorbers, and other methods. In 1999, Kinetic was bought out by Tenneco. Later developments by a Catalan company, Creuat has devised a simpler system design based on single-acting cylinders. After some projects on competition Creuat is active in providing retrofit systems for some vehicle models.

Historically, the first mass production car with front to rear mechanical interconnected suspension was the 1948 Citroën 2CV. The suspension of the 2CV was extremely soft—the longitudinal link was making pitch softer instead of making roll stiffer. It relied on extreme antidive and antisquat geometries to compensate for that. This redunded into a softer axle crossing stiffness that anti-roll bars would have otherwise compromised. The leading arm/trailing arm swinging arm, fore-aft linked suspension system together with inboard front brakes had a much smaller unsprung weight than existing coil spring or leaf designs. The interconnection transmitted some of the force deflecting a front wheel up over a bump, to push the rear wheel down on the same side. When the rear wheel met that bump a moment later, it did the same in reverse, keeping the car level front to rear. The 2CV had a design brief to be able to be driven at speed over a ploughed field. It originally featured friction dampers and tuned mass dampers. Later models had tuned mass dampers at the front with telescopic dampers/shock absorbers front and rear.

The British Motor Corporation was also an early adopter of interconnected suspension. A system dubbed Hydrolastic was introduced in 1962 on the Morris 1100 and went on to be used on a variety of BMC models. Hydrolastic was developed by suspension engineer Alex Moulton and used rubber cones as the springing medium (these were first used on the 1959 Mini) with the suspension units on each side connected to each other by a fluid filled pipe. The fluid transmitted the force of road bumps from one wheel to the other (on the same principle as the Citroen

2CV's mechanical system described above) and because each suspension unit contained valves to restrict the flow of fluid also served as a shock absorber. Moulton went on to develop a replacement for Hydrolastic for BMC's successor, British Leyland. This system manufactured under licence by Dunlop in Coventry, called Hydragas worked on the same principle but instead of rubber spring units it used metal spheres divided internally by a rubber diaphragm. The top half contained pressurised gas and the lower half the same fluid as used on the Hydrolastic system. The fluid transmitted suspension forces between the units on each side whilst the gas acted as the springing medium via the diaphragm. This is the same principle as the Citroen hydropneumatic system and provides a similar ride quality but is self-contained and does not require an engine-driven pump to provide hydraulic pressure. The downside is that Hydragas is, unlike the Citroen system, not height adjustable or self-levelling. Hydragas was introduced in 1973 on the Austin Allegro and was used on several models, the last car to use it being the MG F in 2002. The system was changed in favour of coil springs over dampers, due to cost reasons, towards the end of the vehicle's life. When it was decommissioned in 2006 the Hydragas manufacturing line was over 40 years old.

Some of the last post-war Packard models also featured interconnected suspension.

Types

- Live axle with a Watt's link.
- Sliding pillar.
- Swing axle.
- Double wishbone suspension.
- MacPherson.

This diagram is not exhaustive; notably excluding elements such as trailing arm links and those that are flexible.

Suspension systems can be broadly classified into two subgroups: dependent and independent. These terms refer to the ability of opposite wheels to move independently of each other.

A dependent suspension normally has a beam (a simple 'cart' axle) or (driven) live axle that holds wheels parallel to each other and perpendicular to the axle. When the camber of one wheel changes, the camber of the opposite wheel changes in the same way (by convention on one side this is a positive change in camber and on the other side this a negative change). De Dion suspensions are also in this category as they rigidly connect the wheels together.

An independent suspension allows wheels to rise and fall on their own without affecting the opposite wheel. Suspensions with other devices, such as sway bars that link the wheels in some way are still classed as independent.

A third type is a semi-dependent suspension. In this case, the motion of one wheel does affect the position of the other but they are not rigidly attached to each other. A twist-beam rear suspension is such a system.

Dependent Suspensions

Dependent systems may be differentiated by the system of linkages used to locate them, both longitudinally and transversely. Often both functions are combined in a set of linkages.

Examples of location linkages include:

- Satchell link.
- Panhard rod.
- Watt's linkage.
- WOBLink.
- Mumford linkage.
- Leaf springs used for location (transverse or longitudinal).
  - Fully elliptical springs usually need supplementary location links and are no longer in common use.
  - Longitudinal semi-elliptical springs used to be common and still are used in heavy-duty trucks and aircraft. They have the advantage that the spring rate can easily be made progressive (non-linear).
  - A single transverse leaf spring for both front wheels and both back wheels, supporting solid axles, was used by Ford Motor Company, before and soon after World War II, even on expensive models. It had the advantages of simplicity and low unsprung weight (compared to other solid axle designs).

In a front engine, rear-drive vehicle, dependent rear suspension is either "live axle" or deDion axle, depending on whether or not the differential is carried on the axle. Live axle is simpler but the unsprung weight contributes to wheel bounce.

Because it assures constant camber, dependent (and semi-independent) suspension is most common on vehicles that need to carry large loads as a proportion of the vehicle weight, that have relatively soft springs and that do not (for cost and simplicity reasons) use active suspensions. The use of dependent front suspension has become limited to heavier commercial vehicles.

Independent Suspensions

The variety of independent systems is greater and includes:

- Swing axle.
- Sliding pillar.
- MacPherson strut/Chapman strut.
- Upper and lower A-arm (double wishbone).
- Multi-link suspension.
- Semi-trailing arm suspension.
- Swinging arm.

- ○ Transverse leaf springs when used as a suspension link, or four quarter elliptics on one end of a car are similar to wishbones in geometry, but are more compliant. Examples are the front of the original Fiat 500, the Panhard Dyna Z and the early examples of Peugeot 403 and the back of the AC Ace and AC Aceca.

Because the wheels are not constrained to remain perpendicular to a flat road surface in turning, braking and varying load conditions, control of the wheel camber is an important issue. Swinging arm was common in small cars that were sprung softly and could carry large loads, because the camber is independent of load. Some active and semi-active suspensions maintain the ride height, and therefore the camber, independent of load. In sports cars, optimal camber change when turning is more important.

Wishbone and multi-link allow the engineer more control over the geometry, to arrive at the best compromise, than swing axle, MacPherson strut or swinging arm do; however the cost and space requirements may be greater. Semi-trailing arm is in between, being a variable compromise between the geometries of swinging arm and swing axle.

Semi-independent Suspension

In a semi-independent suspensions, the wheels of an axle are able to move relative to one another as in an independent suspension but the position of one wheel has an effect on the position and attitude of the other wheel. This effect is achieved via the twisting or deflecting of suspension parts under load. The most common type of semi-independent suspension is the twist beam.

SECURITY SYSTEMS

Motor vehicle theft has been a problem since the start of the automobile age. The 1900 Leach automobile featured a removable steering wheel that the driver could carry away to prevent unauthorized vehicle use. More recently, sophisticated electronic alarms, some of which incorporate radio beacons, and more tamper-resistant wiring and electronic locks have been produced. Through the use of wireless technology, vehicles equipped with Global Positioning System (GPS) satellite navigation systems may be tracked and recovered when stolen.

Car and alarm manufacturers are constantly trying to improve security against theft. Great achievements have been made recently by incorporating the alarm system as an integral part of the vehicle electronics. Even the retro-fit systems are still very effective. Three main types of intruder alarm are used:

- Switch operated on all entry points.
- Battery voltage sensed.
- Volumetric sensed.

In addition the methods used to disable the vehicle are : (a) Ignition and starter circuit cut off.(b) Engine ECU code lock.

Alarms can be set by a separate switch or IR transmitter. More commonly now, they are set automatically when the doors are locked. Professional car thieves always find ways to maneuver the latest alarm systems. The vehicle manufacturers strive to stay one step ahead. Legislation is being considered for installing tracking devices in an unknown part of a vehicle's chassis. This can be activated during the theft of the car, allowing the police to trace the vehicle.

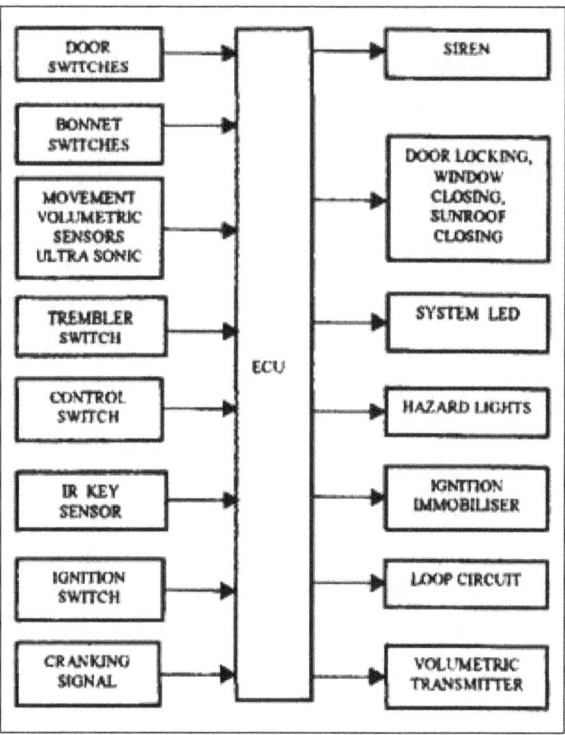

Block diagram of a complex alarm system.

The good alarm systems now available are either retro-fit systems or factory fitted. Most are made for 12 V negative earth vehicles. They use electronic sirens and generate an audible signal when arming and disarming. All of them are triggered when the car door opens and automatically re-set after a period of time, often 1 or 2 minutes. The alarms are triggered instantly when the entry point is breached. Most systems are two pieces, with separate control unit and siren. The control unit is installed in the passenger compartment and the siren under the bonnet in most systems. Most recent systems use two infra-red remote keys, which incorporate small button type batteries and have an LED to indicate when the signal is being sent. They operate with one vehicle only. Intrusion sensors, which detect car movement or use volumetric sensing, can be adjusted for sensitivity. When operating with flashing lights most systems draw current about 5 A and without flashing lights (siren only) the current drawn is less than 1 A. The sirens produce a sound level of about 95 dB, when measured 2 m in front of the vehicle. A block diagram of a complex alarm system is shown is Fig. The system, as usual, contains a series of inputs and outputs as follows.

Inputs:

- Ignition supply.
- Engine cranks signal.
- Volumetric sensor.
- Bonnet switch.

- Tremble switch.
- IR/RF remote signal.
- Door switches.
- Control switch.

Outputs:

- Volumetric transmitter.
- System LED.
- Horn or siren.
- Hazard lights.
- Ignition immobilizer.
- Loop circuit.
- Electric windows, sun roof and door locks.

Some factory fitted alarms are coupled with the central door locking system, known as lazy lock. One press of the remote unit sets the alarm, close windows and sun roof and finally locks the doors.

Security Coded Electronic Control Units

One of the recent ideas is to use a security code in engine electronic control unit so that it can be unlocked to allow the engine to start only when it receives a coded signal. Ford is using a special ignition key, programmed with the required information. Citroen uses a similar idea in some of their models, but the code has to be entered via a numerical keypad. Of course with this arrangement also the car can be lifted onto a lorry and driven away, but when this is done a new engine control ECU is required, which is costly.

SAFETY SYSTEMS

Picture of a crash test performed at General Motors.

Vehicle Safety Technology (VST) in the automotive industry refers to the special technology developed to ensure the safety and security of automobiles and their passengers. The term encompasses a broad umbrella of projects and devices within the automotive world. Notable examples of VST include geo-fencing capabilities, remote speed sensing, theft deterrence, damage mitigation, vehicle-to-vehicle communication, and car-to-computer communication devices which use GPS tracking.

Preventative Technologies for Driver Drowsiness

To prevent or account for drowsiness while driving, many companies have researched technologies to either detect drowsiness and protect the driver or keep the driver awake. One technology that aims to keep drivers awake is blue light. Blue light strains the eyes, making it difficult to fall asleep while driving, and it may be contained in all interior lights, touchscreens, clocks, and lights illuminating the speedometer and gas meter. Another preventative technology, designed to detect drowsiness, works by using data from lane departure sensors to identify jerky movements or swerving in and out of lanes. Once received, a coffee symbol on the dashboard will illuminate or a verbal response will sound to alert the driver that they should take a break. The driver's seat can also vibrate to startle them in the hopes that they become more alert. Both of these technologies are not yet perfect and are often faulty in detecting drowsiness. Other technologies are also being developed, including the flashing of bright lights containing blue light to keep drivers awake, along with steering technology that can correct for driver error while swerving due to drowsiness. Other technology seeks to prevent accidents from occurring by analyzing driver behavior. Companies such as SafeMode operate by analyzing driver behavior for "safety events", such as hard accelerations or breaking. If a driver decreases the amount of times they record a safety event month after month, they earn incentives, ultimately making the road safer for all drivers.

Electronic Stability Control (ESC)

Electronic stability control (also known as roll over protection) is a specific technology that helps keep the vehicle balanced. During harsh weather or tough road conditions that would cause vehicle steering to be extreme, this technology allows the drivers to regain control and prevent possible crashes, roll overs, and fishtails. This is the system which allows drivers to exit hydroplanes safely. In combination with automatic emergency braking technologies, ESC controls each wheel individually to allow the driver to steer in the intended direction. To do this, ESC technologies apply singular breaks to all the wheels individually, slowing each one to the intended speed of the rest. Electronic stability control does not give traction to the vehicle; rather, it provides balance and momentary steering control.

Automatic and Emergency Braking

There are four different and diverse automatic and emergency braking technologies. They include Automatic Emergency Braking (AEB), Crash Imminent Braking, Dynamic Brake Support, and Pedestrian Emergency Braking. These collision avoidance technologies detect vehicles in front of the car and automatically brake if a crash is detected. Before making any decisions on their own, AEB systems alert the driver of the suspected crash and allow a chance for the driver to take action. If the driver does not acknowledge the alert, AEB technology will then apply the brakes in hopes of avoiding or lessening the severity of a crash. Dynamic brake support is a technology which supplements the driver's brake if it is not hard enough already. As for crash imminent braking, this system automatically applies the brakes to avoid a crash. Drivers do not need to apply the brakes for the crash imminent system to engage. Pedestrian emergency braking systems sense pedestrians in front or near the car and will apply the brakes if drivers make no move to do so. This technology uses the front facing radar sensors and cameras to detect pedestrians, then apply the brakes in hopes of avoiding a collision. These systems have been available on a wide range of vehicles since 2006. Due to their growth in popularity, prices to add these systems to new vehicles have dropped and they are inexpensive to install. However they are quite pricey to fix due to the resolution and grade of cameras and sensors.

Blind Spot Monitoring

The vehicle blind spot is an area outside the vehicle which cannot be seen by the driver from the driver's seat. Each person will have a different blind spot, and cannot see other cars within that area. To reduce the occurrence of crashes related to blind spots, numerous companies have developed technologies that alert drivers to other cars near their vehicle. These technologies also have the ability to detect which side other vehicles are approaching from. Vehicles with blind spot monitoring employ radar sensors and cameras positioned all around the exterior which warn drivers of the movement and presence of other vehicles in their surroundings. To alert the driver, blind spot monitoring systems will use one or more of warning sounds, seat vibration, and illuminated warning symbols, usually located on the outside of the vehicle's side mirrors on the side where a vehicle is present. Noises may also sound if the driver engages the turn signal when there is a vehicle in the blind spot area. Blind spot monitoring technologies are equipped on most luxury vehicles and in recently produced vehicles. There are packages present to add blind spot monitoring to vehicles at the time of purchase. One inherent weakness of many of these systems is that they struggle to detect fast moving cars, motorcycles, or low riding cars.

Lane Departure Warnings

Lane departure warning systems are technologies that employ underside and wheel well cameras to detect when a wheel has unintentionally crossed a lane line without a turn signal. These systems were created in order to prevent lane swerving and possible driver drowsiness accidents. To alert the driver of a lane departure, systems can do employ one or more of a seat vibration, the illumination of a warning signal, and warning sounds (either a small sound or a verbal warning from system technologies. These systems do not alert the driver of a lane departure if a turn signal is used. These systems intend to prevent crashes and drowsiness by reducing the number of times a lane switch occurs without the use of a turn signal. Lane departure systems can be disabled so that no alert will occur, but it is recommended that a dealer make this change to prevent damage to the system.

Speed Monitoring and Warning Systems

Speed warning systems are designed to alert the driver of the vehicle when they have exceeded the speed limit. To do this, GPS technologies are used to triangulate the vehicle's location; along with a record of speed limits in the area, the system uses built-in speed sensors to notify the driver when they exceed the speed limit. If the list of speed limits is updated, it can also track school and work zone speed limit changes that would otherwise go unnoticed by the driver. When a driver exceeds the speed limit, any of the following can occur: a small ding will sound, the speed reading will turn red if the odometer is digital, an icon will light up and a ding will sound, or the vehicle will give a verbal response alerting the driver of their speed and to slow down. One speed monitoring technology which is undergoing testing in the US and Europe is called intelligent speed adaptation, which would assess the speed of the vehicle, and with the permission from the driver, automatically slow the vehicle down to the proper legal speed. If implemented, though, drivers would be able to override this system or turn it off entirely. However, these technologies keep drivers safe from distractions which would otherwise result in a speeding ticket or a car accident.

IGNITION SYSTEM

This ignition takes place thanks to a group of components working together, otherwise known as the ignition system. The ignition system consists of an ignition coil, distributor, distributor cap, rotor, plug wires and spark plugs. Older systems used a points-and-condenser system in the distributor, newer (as in most we'll ever see anymore) use an ECU, a little brain in a box, to control the spark and make slight changes in ignition timing.

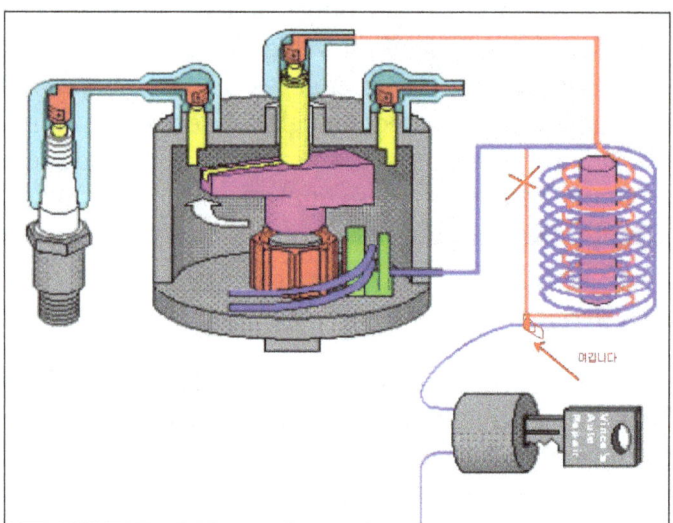

This diagram shows the parts of your ignition system.

The Ignition Coil

The ignition coil is the unit that takes your relatively weak battery power and turns it into a spark powerful enough to ignite fuel vapor. Inside a traditional ignition coil are two coils of wire on top of each other. These coils are called windings. One winding is called the primary winding, the other is the secondary. The primary winding gets the juice together to make a spark and the secondary sends it out the door to the distributor.

You'll see three contacts on an ignition coil unless it has an external plug, in which case the contacts are hidden inside the case. The large contact in the middle is where the coil wire goes (the wire that links the coil to the distributor cap. There is also a 12V+ wire that connects to a positive power source. The third contact communicates information to the rest of the car, like the tachometer.

The Distributor, Distributor Cap and Rotor

Once the coil generates that very powerful spark, it needs to send it someplace. That someplace takes the spark and sends it out to the spark plugs, and that someplace is the distributor.

The distributor is basically a very precise spinner. As it spins, it distributes the sparks to the individual spark plugs at exactly the right time. It distributes the sparks by taking the powerful spark that came in via the coil wire and sending it through a spinning electrical contact known as the rotor. The rotor spins because it's connected directly to the shaft of the distributor. As the rotor spins, it makes contact with a number of points (4, 6, 8 or 12 depending on how many cylinders your engine has) and sends the spark through that point to the plug wire on the other end. Modern distributors have electronic assistance that can do things like alter the ignition timing.

Spark Plugs and Wires

After the coil takes the weaker juice and makes a high powered spark and the distributor takes the powerful spark and spins it to the right outlet, we need a way to take the spark to the spark plug. This is done through the spark plug wires. Each contact point on the distributor cap is connected to a plug wire that takes the spark to the spark plug.

The spark plugs are screwed into the cylinder head, which means that the end of the plug is sitting at the top of the cylinder where the action happens. At just the right time (thanks to the distributor), when the intake valves have let the right amount of fuel vapor and air into the cylinder, the spark plug makes a nice, blue, hot spark that ignites the mixture and creates combustion.

At this point, the ignition system has done its job, a job it can do thousands of times per minute.

The Ignition Module

In the old days, a distributor relied on a lot of its own "mechanical intuition" to keep the spark timed perfectly. It did this through a setup called a points-and-condenser system. Ignition points were set to a specific gap that created an optimal spark while the condenser regulated.

These days this is all handled by computers. The computer that directly regulates your ignition system is called the ignition module, or ignition control module. There is no maintenance or repair procedure for the module aside from replacement.

EXHAUST SYSTEM

The exhaust system collects the exhaust gases from the cylinders, removes harmful substances, reduces the level of noise and discharges the purified exhaust gases at a suitable point of the vehicle away from its occupants.

Components of an Exhaust – Pipes

The exhaust system is comprised mainly of pipes in several different shapes, each designed to connect to one another, and each shaped to conform to a specific part of the underside of the car. (The pipes are often bent to wrap around or otherwise accommodate other nearby components of the car, such as the axles.) Each pipe is responsible for moving the exhaust gases toward the back, but many of the segments are specialized. In other words, from the exterior, the system simply looks like a bunch of connected pipe segments that run from the engine bay to the back bumper, though some serve an additional purpose as the exhaust flows through that particular pipe.

For example, the Y pipe (which is simply, well, a Y-shaped pipe) might be installed so the end with two openings bolts up to two corresponding openings in the exhaust manifold, combining the engine's waste to progress through the system. Or, when installed at the back end of the car, a Y pipe can help create a dual exhaust system with a tailpipe on each side of the car (for an often sought-after sporty look). Intermediate pipes might be attached to the muffler or resonator, which are other important system components. Balance pipes, found in dual exhaust systems, help equalize the exhaust pulses traveling under the driver and passenger sides of the car. Air gap pipes are specialized nested pipes that act as a heat shield and insulator by providing an extra layer for airflow. And the tail pipe, which is typically peeking out from underneath the rear bumper, usually has a larger opening and might be made of more substantial-looking metal, to give the appearance of a performance exhaust that's a common feature of high-end cars.

Though it might seem inefficient to have a ton of pipes instead of just one, really, all those segments serve a purpose. For one, bending pipes is hard work, and it's easier to connect small angled segments to straight pipes than it is to shape one long, expensive, heavy pipe to fit every contour of a car. Also, exhaust system components wear out at regular intervals (depending, of course, on the manufacturer, its materials, driving conditions and environmental factors). It's easier and less expensive to replace one rusted-out segment of pipe, banged-up muffler, or worn-out catalytic converter than it would be to install a whole new system.

Other Exhaust Components – Headers, Hangers and more

the exhaust system really starts at the exhaust header (also known as the exhaust manifold). The header looks like a series of adjacent tubes stuck together (though it's often made of heavy-duty cast metal). It collects the exhaust directly from the engine, so it's designed so each opening mates up with one of the engine's exhaust ports, with flanges that form a tight seal to prevent exhaust from escaping. From here, the exhaust begins its flow through the various pipes and other components.

The gaskets that are fitted between each pipe play a very important role. It's difficult, if not impossible, to form a flawless metal-to-metal seal, so gaskets are sandwiched between each connection to prevent the poisonous gases from escaping prematurely. They're made of fiber or other heavy-duty, heat resistant materials, which have just enough flexibility to compress slightly when the pipes are clamped tightly together. This helps form a tight seal.

The muffler is another key part of your car's exhaust system.

The muffler is another key part of the exhaust system. It looks like a large round or oval chamber (usually, but not always, found near the back of the car). That chamber has a very complex design, though – it's responsible for silencing most of the engine's noise, even though it has to allow the exhaust to continue flowing smoothly. A series of chambers and tubes, filled with rock, wool, or synthetic fibers, absorbs and controls the noise. And that's not all – the muffler must be reasonably resistant to damage, corrosion and heat retention. The resonator is a secondary or substitute sound elimination component, used to augment the muffler, or in cases where space is at a premium it might even be used instead of a muffler.

Catalytic converters are the primary and most efficient means of reducing the level of toxins in a car's exhaust. The innards of a catalytic converter (or "cat") are coated with metals. Different types of catalytic converters use different combinations of platinum, palladium and rhodium. Before the exhaust reaches the cat, it contains a potent, super-toxic combination of carbon monoxide, nitrogen oxide and hydrocarbons. (When you take your car for its emissions test, these are the chemical levels that are being tested.) When these poisons come into contact with the metals coating the inside of the cat, a chemical reaction takes place that makes the exhaust gases less harmful. As the exhaust passes through the cat, the level of chemicals should be reduced enough to comply with government regulations.

We already talked about gaskets, but the exhaust system requires other basic pieces of hardware. Flanges generally serve the same purpose as gaskets, but these are made of metal (and are sometimes formed right onto the end of a pipe). Assorted clamps and brackets mount the exhaust pieces together and hold them to the vehicle, and exhaust hangers literally hang the pipes from the underbody of the car, with enough strength to keep them in place but also enough flexibility to withstand movement caused by driving. And last, but not least are the heat shields: Metal (sometimes insulated) plates that are used as an extra barrier whenever hot exhaust parts are particularly close to another part of the car or directly below the passenger compartment.

The Exit Route

If an engine ran perfectly, it would combust all its available fuel as it ran through its cycles, converting all the dirty bits to a source of power. However, leftovers exist as pollutants because an engine simply cannot be engineered to run perfectly – there are far too many variables. Some amount of fuel will always remain unburned or partially burned, and these remnants must be quickly processed out of the vehicle, in the form of exhaust, to make room for the next cycle of engine combustion.

the exhaust first exits the engine and enters the system through the exhaust manifold. From there, it travels down the system through interconnected pipes until it exits through the tailpipe, near the back bumper. The pipes themselves actually help cool the exhaust, but they're mostly a way for the exhaust to travel to (and through) the catalytic converter and muffler.

The cat has to be as close to the engine as possible, because it isn't fully functional until it rises to operating temperature. In many cases, the manufacturer places the cat shortly after the manifold, so heat from the engine helps warm the cat and quickly bring it up to temp.

After the gases pass through the cat, which will burn off and remove up to 90 percent of the exhaust's toxins, the next priority is to filter the engine sound. The muffler and resonator are usually situated right beyond the cat. There are many variations on this combination – some will soothe the exhaust as much as possible, while others are specifically tuned for aggressive tones. From there, the exhaust moves through the remainder of the pipes until it exits the car.

With all those chemicals swirling around, it's quite a feat that the exhaust system actually works as well as it does. A well-maintained exhaust system should last two to three years, but the pipes incur damage both inside and out. On the outside, they're susceptible to road conditions, such as impact from debris and environmental factors, such as snow, ice and road salt.

However, a more substantial cause of exhaust system degradation is internal, and it can't be seen until the pipes have corroded through with rust. We know that as the engine combusts fuel to make power, byproducts of this process are left over – that's why the exhaust system is needed. One of these byproducts is acidic moisture, and it's really damaging to metal. Unfortunately, there's really no way to keep the insides of the pipes clean.

When an exhaust pipe rots through or a connection comes loose, an exhaust leak occurs. A leak is almost always immediately apparent with a loud, obnoxious sound and, possibly, drivability issues like an intermittent fluctuation in power. But, believe it or not, these aren't the real problems. It gets a little more serious if the leak occurs before the catalytic converter, which means the exhaust isn't being properly processed and all those hot chemicals are spilling everywhere into the atmosphere.

Keeping the Pipes Clean

Diesel engines, often blamed for disproportionately high levels of toxic emissions, may find their reputation salvaged somewhat by new and improved exhaust filters that claim to reduce emissions by about 25 percent, with the added benefit of increasing fuel economy. Several automotive manufacturers with a history of offering diesel models have introduced similar systems that combine high-powered catalytic converters with diesel particulate filters. Volkswagen's BlueMotion turbodiesel lineup (sold mostly in Europe and South America) and Mercedes' BlueTec diesels are two such examples. In many cases, these OEM components and car models use weight reduction and other streamlining techniques to improve overall efficiency.

Those chrome exhaust tips aren't really fooling anyone.

Improved exhaust efficiency is critical to getting the most out of your car, and there are even ways to improve an older car (assuming it's in generally good condition – a performance exhaust system won't work miracles on a total jalopy). A mechanic can usually order identical original replacement parts, but there are also alternatives from aftermarket performance parts suppliers that often give the car a facelift (at least, as much as the car's underbody can be spruced up, anyway). There are several frequently-mentioned benefits of performance exhaust systems. Aftermarket manufacturers are well aware that the auditory experience is critical to a pleasurable drive, and it also lets nearby drivers know that a little extra coin was spent on the car, too. Most manufacturers offer a variety of systems that range in sound from a subtle throatiness to an all-out roar. A good aftermarket performance exhaust can improve throttle response (how fast, and smoothly, the car reacts to the pressure applied to the gas pedal) and give a boost in horsepower. Some systems can even improve fuel economy. Performance exhausts are able to provide this power boost because they're less restrictive than stock exhausts. If the engine can push out more air (which also means it can take in more air) that helps the engine make more power. But contrary to what you might think, a totally free-flowing system will be counterproductive. Aftermarket exhausts are engineered to provide just the right amount of back pressure so that the engine doesn't end up running at sub-peak output. That's why it's important to choose one designed and optimized for your specific vehicle. And before you go ripping out your exhaust pipes, make sure the system you want is street-legal and won't cause your car to fail its emissions test.

Understand, too, that a real performance exhaust system cannot provide mind-blowing benefits for a trivial investment. Beware of cheap products like clip-on or screw-on "performance" exhaust system components or exhaust tips that claim to give the look of a full performance system. They won't help the car's drivability and they may even hinder it by adding extra weight and wind resistance.

References

- Engine-lubrication, automotive-lubrication, knowledge-center, en: masterlineworld.com, Retrieved 22 January, 2019

- "Trends in the Semiconductor Industry: 1970s". Semiconductor History Museum of Japan. Retrieved 27 June 2019

- How-The-Lubrication-System-Works-In-An-Engine, news: lubrita.com, Retrieved 23 February, 2019

- Scrosati, Bruno; Garche, Jurgen; Tillmetz, Werner (2015). Advances in Battery Technologies for Electric Vehicles. Woodhead Publishing. ISBN 9781782423980

- Lubrication-systems-for-petrol-engines-automobile, automobile: what-when-how.com, Retrieved 24 March, 2019

- "Suspension Basics 9 - Hydropneumatic Springs". Initial Dave. Archived from the original on 2015-01-29. Retrieved 2015-01-29

- Coolingsystem, classroom: carparts.com, Retrieved 25 April, 2019

- Development-in-transmission-system, useful-information: automobiles.mapsofindia.com, Retrieved 26 May, 2019

- How-the-braking-system-works, basics: howacarworks.com, Retrieved 27 June, 2019

Automobile Safety

Automobile safety deals with the study of design, construction and regulations for minimizing the occurrence and consequences of traffic accidents. Crashworthiness, global NCAP crash test, roll-over, active safety, etc. are some concepts that fall within it. This chapter discusses these concepts of automobile safety in detail.

Automobile safety is the study and practice of design, construction, equipment and regulation to minimize the occurrence and consequences of automobile accidents. Road traffic safety more broadly includes roadway design. One of the first formal academic studies into improving vehicle safety was by Cornell Aeronautical Laboratory of Buffalo, New York. The main conclusion of their extensive report is the crucial importance of seat belts and padded dashboards. However, the primary vector of traffic-related deaths and injuries is the disproportionate mass and velocity of an automobile compared to that of the predominant victim, the pedestrian.

In the United States a pedestrian is injured by an automobile every 8 minutes, and are 1.5 times more likely than a vehicle's occupants to be killed in an automobile crash per outing. Improvements in roadway and automobile designs have steadily reduced injury and death rates in all first world countries. Nevertheless, auto collisions are the leading cause of injury-related deaths, an estimated total of 1.2 million in 2004, or 25% of the total from all causes. Of those killed by autos, nearly two-thirds are pedestrians. Risk compensationtheory has been used in arguments against safety devices, regulations and modifications of vehicles despite.

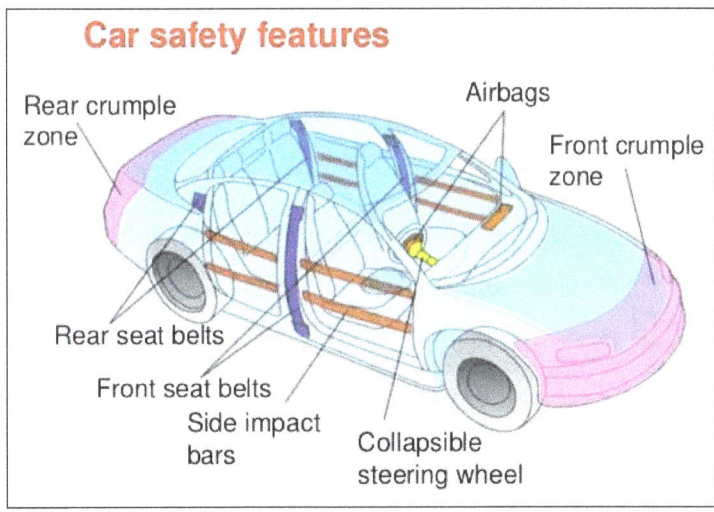

In order to understand how seat belts and air bags provide protection it is important to know about occupant movement in the event of an accident. Upon impact the vehicle either decelerated rapidly or accelerated when hit by a moving object such as another vehicle. When this occurs inertia always causes the occupants to move in the opposite direction to the applied impact force. The safety systems are intended to reduce the risk of contact between the occupants and vehicle interior. However glasses generally prevent the risk of injuries caused by objects entering the vehicle from outside.

ETD and Seat Belt Force Limiter

The front seat belts and some of the outer rear belts are equipped with ETDs. In addition in some seat belt systems, belt force limiters are included. If deployed in event of an accident, an ETD reduces these at belt slack. The occupant is restrained earlier and thus participates sooner in the vehicle's deceleration decreasing the occupant load during the collision. In addition seat belt is also equipped with a force limiter it reduces upon activation the peak seat belt force excreted on the occupant. The belt force limiter is tuned to the front air bag which in turn takes on some of the seat belt induced forces thus providing a more even load distribution. When the ignition is on the ETD s are deployed during a frontal or rear end collision of sufficient severity that is accidents with high longitudinal deceleration or acceleration sensed and made more safer.

Active and Passive Safety

The terms "active" and "passive" are simple but important terms in the world of automotive safety. "Active safety" is used to refer to technology assisting in the prevention of a crash and "passive safety" to components of the vehicle (primarily airbags, seatbelts and the physical structure of the vehicle) that help to protect occupants during a crash.

Crash Avoidance

Crash avoidance systems and devices help the driver — and, increasingly, help the vehicle itself — to avoid a collision. This category includes:

- The vehicle's headlamps, reflectors, and other lights and signals the vehicle's mirrors.

- The vehicle's brakes, steering, and suspension systems.

Driver Assistance

A subset of crash avoidance is driver assistance systems, which help the driver to detect obstacles and to control the vehicle. Driver assistance systems include:

- DADS: Driver Alertness Detection System. System to prevent accident caused by fatigue.

- Automatic Braking systems to prevent or reduce the severity of collision.

- Infrared night vision systems to increase seeing distance beyond headlamp range.

- Adaptive headlamps control the direction and range of the headlight beams to light the driver's way through curves and maximize seeing distance without partially blinding other drivers.

- Reverse backup sensors, which alert drivers to difficult-to-see objects in their path when reversing.

Backup Camera

Lights and Reflectors

Vehicles are equipped with a variety of lights and reflectors to mark their presence, position, width, length, and direction of travel as well as to convey the driver's intent and actions to other drivers. These include the vehicle's headlamps, front and rear position lamps, side marker lights and reflectors, turn signals, stop (brake) lamps, and reversing lamps. School buses and Semi-trailer trucks in North America are required to bear retro reflective strips outlining their side and rear perimeters for greater conspicuity at night.

Unused Safety Features

Many different inventions and ideas which may or may not have been practical about auto safety have been put forward but never made it to a production car. Such items include the driver seat in the middle (to give the person a better view) (the exception being the Mclaren F1 sports car), rear facing seats (except for infant car seats), and control stick steering.

Post-crash Survivability

Post-crash survivability is the chance that drivers and passengers survive a crash after it occurs. Technology such as Advanced Automatic Collision Notification can automatically place calls to emergency services and send information about a vehicle collision.

Infants and Children

Children present significant challenges in engineering and producing safe vehicles, because most children are significantly smaller and lighter than most adults. Safety devices and systems designed and optimized to protect adults — particularly calibration-sensitive devices like airbags and active seat belts — can be ineffective or hazardous to children. In recognition of this, many medical professionals and jurisdictions recommend or require that children under a particular age, height, and weight ride in a child seatand in the back seat, as applicable. In Sweden, for instance, a child or an adult shorter than 140 cm is legally forbidden to ride in a place with an active airbag in front of it. Child and driver-controlled power window lockout controls prevent children from opening doors and windows from inside the vehicle.

ACTIVE SAFETY

The term active safety (or primary safety) is used in two distinct ways.

The first, mainly in the United States, refers to automobile safety systems that help avoid accidents, such as good steering and brakes. In this context, passive safety refers to features that help reduce the effects of an accident, such as seat belts, airbags and strong body structures. This use is essentially interchangeable with the terms primary and secondary safety that tend to be used worldwide in standard UK English. The correct ISO term is "primary safety".

However, active safety is increasingly being used to describe systems that use an understanding of the state of the vehicle to both avoid and minimise the effects of a crash. These include braking systems, like brake assist, traction control systems and electronic stability control systems, that interpret signals from various sensors to help the driver control the vehicle. Additionally, forward-looking, sensor-based systems such as advanced driver-assistance systems including adaptive cruise control and collision warning/avoidance/mitigation systems are also considered as active safety systems under this definition.

These forward-looking technologies are expected to play an increasing role in collision avoidance and mitigation in the future. Most major component suppliers, such as Aptiv, TRW and Bosch, are developing such systems. However, as they become more sophisticated, questions will need to be addressed regarding driver autonomy and at what point these systems should intervene if they believe a crash is likely.

In engineering, active safety systems are systems activated in response to a safety problem or abnormal event. Such systems may be activated by a human operator, automatically by a computer driven system, or even mechanically. In nuclear engineering, active safety contrasts to passive safety in that it relies on operator or computer automated intervention, whereas passive safety systems rely on the laws of nature to make the reactor respond to dangerous events in a favourable manner.

Examples:

- The computer operated control rods in a nuclear power station provide an active safety system, whereas a fuel that produces less heat at abnormally high temperatures constitutes a passive safety feature.

- Collision avoidance systems in a modern car.

- Many buildings have interconnected fire alarms that can be triggered manually by pushing a button or breaking a glass plate attached to sensors.

Automotive Sector

In the automotive sector the term active safety (or primary safety) refers to safety systems that are active *prior* to an accident. This has traditionally referred to non-complex systems such as good visibility from the vehicle and low interior noise levels. Nowadays, however, this area contains highly advanced systems such as anti-lock braking system, electronic stability control and collision warning/avoidance through automatic braking. This compares with passive safety (or secondary safety), which are active *during* an accident. To this category belong seat belts, deformation zones and airbags, etc.

Advancement in passive safety systems has progressed very far over the years, and the automotive industry has shifted its attention to active safety where there are still a lot of new unexplored areas. Research today focuses primarily on collision avoidance (with other vehicles, pedestrians and wild animals) and vehicle platooning.

Examples of Active Safety

- Good visibility from driver's seat.

- Low noise level in interior.

- Legibility of instrumentation and warning symbols.

- Early warning of severe braking ahead.

- Head up displays.

- Good chassis balance and handling.

- Good grip.

- Anti-lock braking system.

- Electronic Stability Control.

- Chassis assist.

- Intelligent speed adaptation.

- Brake assist.

- Traction control.

- Collision warning/avoidance.

- Adaptive or autonomous cruise control system.

- Electronic brakeforce distribution.

Examples of Passive Safety

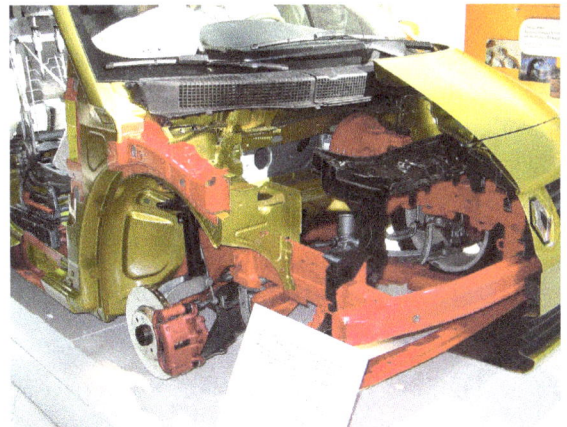

Front structure of a Renault Scénic showing crumple zones.

- Passenger safety cell.

- Crumple zones.

- Seat belts.

- Loadspace barrier-nets.

- Air bags.

- Laminated glass.

- Correctly positioned fuel tanks.

- Fuel pump kill switches.

SIDE COLLISION

Side collisions are vehicle crashes where the side of one or more vehicles is impacted. These crashes often occur at intersections, in parking lots, and when two vehicles pass on a multi-lane roadway.

Occurrences and Effects

For fatalities, in the United States, in 2008, a total of 5,265 (22%) out of 23,888 people were killed in vehicles which were struck in the side.

For speed, in Europe in 2015, it is considered that best designed cars provide serious front crash protection with speeds up to 70 km/h for car occupants wearing seat belts in frontal impacts and 50 km/h in side impacts It is considered that passenger car fatalities and seriously injured side impacts account for about 35 to 40%. In most European countries, another stakeholder is involved in the side impact, with a rate between 45% and 66%. But side impact (22% to 29%) is less common that frontal impact (61% to 69%).

For European motorcyclists, side impact is the second most frequent location of impact.

For European cyclist, thorax injuries are associated with side-impact injuries in urban areas and at junctions.

In European countries, such as UK, Sweden and France, around one quarter of traffic injuries are produced by side collision, but among people subject to killing injuries the side impact account for 29 to 38% of those fatal injuries.

In European vehicle side impact, 60% of casualties were "struck side", while 40% were "non struck side", in 2018.

Fatal casualties count as 50% and 67% in UK and in France, in 2010

Also, side collision are not well managed with child restraints which are not enough taking into account the movement of the child's head and prevent contact with the car's interior.

For light vans and minibuses in 2000 in UK and Germany, between 14% and 26% of accidents with passenger cars are side impact.

In Shanghai, in China, 23% of the 1097 serious accidents occurred between June 2005 and March 2013 are side impact accidents, there the leading collision mode, according to the Shanghai United

Road Traffic Safety Scientific Research Center (SHUFO) database. The head and neck are involved in around 64% of the casualties.

Broadside or T-bone Collision

Broadside collisions are where the side of one vehicle is impacted by the front or rear of another vehicle, forming a "T". In the United States and Canada this collision type is also known as right-angle collision or T-bone collision; it is also sometimes referred to by the abbreviation "AABS" for "auto accident, broadside". Vehicle damage and occupant injury are more likely to be severe, but severity varies based on the part of the vehicle that is struck, safety features present, the speeds of both vehicles, and vehicle weight and construction.

When a vehicle is hit on the side by another vehicle, the crumple zones of the striking vehicle will absorb some of the kinetic energy of the collision. The crumple zones of the struck vehicle may also absorb some of the collision's energy, particularly if the vehicle is not struck on its passenger compartment. Both vehicles are frequently turned from their original directions of travel. If the collision is severe, the struck vehicle may be spun or rolled over, potentially causing it to strike other vehicles, objects, or pedestrians. After the collision, the involved vehicles may be stuck together by the folding of their parts around each other.

An occupant on the struck side of a vehicle may sustain far more severe injuries than an otherwise similar front or rear collision crash.

Side-impact airbags can protect vehicle occupants during side collisions, but they face the same limitations as other airbags. Additionally, side impact wrecks are more likely to involve multiple individual collisions or sudden speed changes before motion ceases. Since the airbag can only provide protection during the first collision, it may leave occupants unprotected during subsequent collisions in the crash. However, the first collision in a crash typically has the most severe forces, so an effective airbag provides maximum benefit during the most severe portion of a crash.

Broadside collisions are frequently caused by a failure to yield right of way. In the case of collisions in an intersection, the cause is often a result of one vehicle failing to obey traffic signals (fail to stop or running past a red light). As with any crash, increased speed may increase crash severity.

Testing

Euro NCAP, IIHS and NHTSA test side impacts in different ways. As of 2015, they all test vehicle-to-vehicle side impacts, where heavier vehicles have lower fatality rates than lighter vehicles.

NHTSA and EuroNCAP also test the more severe vehicle-into-pole side impacts, where smaller vehicles have the same fatality rate as larger vehicles.

Newer cars have improved safety of front crashes, but side impacts are also deadly; about 9,700 people were killed in side impacts in the US in 2004. Side airbags became mandatory in 2009 in the USA, saving an estimated 1,000 lives per year. Research indicates that the vehicle's underbody is the best place to reinforce structures to reduce intrusion by the pole.

ROLLOVER

A rollover is a type of vehicle crash in which a vehicle tips over onto its side or roof. Rollovers have a higher fatality rate than other types of vehicle collisions.

Dynamics

A rolled over Box truck being handled by fire fighters in Jakarta, Indonesia.

A rollover in southern Italy.

Vehicle rollovers are divided into two categories: tripped and untripped. Tripped rollovers are caused by forces from an external object, such as a curb or a collision with another vehicle. Untripped crashes are the result of steering input, speed, and friction with the ground.

Untripped rollovers occur when cornering forces destabilize the vehicle. As a vehicle rounds a corner, three forces act on it: Tire forces (the centripetal force), inertial effects (the centrifugal force), and gravity. The cornering forces from the tire push the vehicle towards the center of the curve. This force acts at ground level, below the center of mass. The force of inertia acts horizontally through the vehicle's center of mass away from the center of the turn. These two forces make the vehicle roll towards the outside of the curve. The force of the vehicle's weight acts downward through the center of mass in the opposite direction. When the tire and inertial forces are enough to overcome the force of gravity, the vehicle starts to turn over.

The most common type of tripped rollover occurs when a vehicle is sliding sideways, and the tires strike a curb, dig into soft ground, or a similar event occurs that results in a sudden increase in lateral force. The physics are similar to cornering rollovers. In a 2003 report, this was the most common mechanism, accounting for 71% of single-vehicle rollovers.

Another type of tripped rollover occurs due to a collision with another vehicle or object. These occur when the collision causes the vehicle to become unstable, such as when a narrow object causes one side of the vehicle to accelerate upwards, but not the other. Turned down guard rail end sections have been shown to do this. A side impact can accelerate a vehicle sideways. The tires resist the change, and the coupled forces rotate the vehicle. In 1983, crash tests showed that light trucks were prone to rolling over after colliding with certain early designs of guide rail.

A rollover can also occur as a vehicle crosses a ditch or slope. Slopes steeper than 33% (one vertical unit rise or fall per three horizontal units) are called "critical slopes" because they can cause most vehicles to overturn.

A vehicle may roll over for other reasons, such as when hitting a large obstacle with one of its wheels or when maneuvering over uneven terrain.

Vehicles

All vehicles are susceptible to rollovers to various extents. Generally, rollover tendency increases with the height of the center of mass, narrowness of the axle track, steering sensitivity, and increased speed.

The rollover threshold for passenger cars is over 1 g of lateral acceleration. The Tesla Model S has an unusually low rollover risk of 5.7% due to its low center of mass. Light trucks will roll over at lateral accelerations of 0.8 to 1.2 g. Large commercial trucks will roll at lateral accelerations as low as 0.2 g Trucks are more likely to roll over than passenger cars because they usually have taller bodies and higher ground clearance. This raises the center of mass.

SUVs are especially prone to rollover, especially those outfitted with long travel off-road suspensions. The increased suspension height for increased clearance off-road raises the center of mass.

Full-size vans don't usually have off-road suspensions, but their increased body height makes them more prone to tip. Fifteen passenger vans such as the Ford E-Series (at 27.9%), are particularly notorious for rolling over because their height is increased by the heavy-duty suspensions necessary to carry large numbers of people. The rollover tendency is increased when the vehicles are heavily loaded. It is recommended to not load anything on the roof of such vans, and to use drivers experienced or trained in safe operation of the vehicle. In such cases, familiarity with the vehicle's behavior loaded and unloaded, avoiding sudden swerving maneuvers, and reducing speed through tight turns can greatly decrease the rollover risk associated with these vehicles.

Manufacturers of SUVs often post warnings on the driver's sun-visor. Among the vehicles which have received publicity for tendencies to roll over are the Ford Bronco II, Suzuki Samurai, Jeep CJ, Mitsubishi Pajero/Montero, and Isuzu Trooper.

Military vehicles have a much wider wheel track than civilian SUVs, making them more difficult to roll over. However, IEDs in Iraq and Afghanistan cause roll overs not seen by civilian vehicles. The top turret gunner is particularly vulnerable.

A tall passenger coach made US headlines when 14 passengers were killed in New York in 2011. The bus swerved, flipped on its side and hit a pole which split off the top of the vehicle.

Exit

After a rollover, the vehicle may end up lying on its side or roof, often blocking the doors and complicating the escape for the passengers. Large passenger vehicles such as buses, trams, and trolley buses that have doors on one side only usually have one or more methods of using windows for escape in case of a rollover. Some have special windows with handles to pull so that windows can be used as an emergency exit. Some have tools for breaking the windows and making an improvised exit. Some have emergency exit door or hatches in their roofs or on the opposite side of the bus to the usual entry door. Some combine two or more of these escape methods.

Roll Bars and Cages

Rollover crashes are particularly deadly for the occupants of a vehicle when compared to frontal, side, or rear crashes, because in normal passenger vehicles, the roof is likely to collapse in towards the occupants and cause severe head injuries. The use of roll cages in vehicles would make them much safer, but in most passenger vehicles their use would cut cargo and passenger space so much that their use is not practical. The Jeep Wrangler, a vehicle which is short, narrow, and designed to be used on uneven terrain, is unusual in that it comes with a roll bar as standard equipment.

The decline in popularity of convertibles in the US was partly caused by concern about lack of protection in rollover accidents, because most convertibles have no protection beyond the windshield frame. Some convertibles provide rollover protection using two protruding curved bars behind the headrests. Some Mercedes-Benz convertibles have a retractable roll bar which deploys in case of an accident. Race cars almost always have roll cages, since racing is very likely to result in a rollover. In addition, the roll cage's chassis-stiffening effect is usually seen as a benefit to the car.

Warning Signs

At least 4 countries have a unique sign warning of curves and other areas with an increased danger of rollover for trucks and other vehicles with high centers of gravity. These signs may include an advisory safe speed to avoid rolling over.

Canada.

United States (with speed limit number).

United States.

United Kingdom (warning of an adverse camber).

New Zealand.

In the UK, the "adverse camber" plate comes with a warning sign such as "roundabout ahead", "bend ahead", "junction on a bend ahead", or "series of bends ahead".

TRAFFIC COLLISION

A traffic collision, also called a motor vehicle collision (MVC) among other terms, occurs when a vehicle collides with another vehicle, pedestrian, animal, road debris, or other stationary obstruction, such as a tree, pole or building. Traffic collisions often result in injury, death, and property damage.

A number of factors contribute to the risk of collisions, including vehicle design, speed of operation, road design, road environment, and driving skills, impairment due to alcohol or drugs, and behavior, notably distracted driving, speeding and street racing. Worldwide, motor vehicle collisions lead to death and disability as well as financial costs to both society and the individuals involved.

In 2013, 54 million people worldwide sustained injuries from traffic collisions. This resulted in 1.4 million deaths in 2013, up from 1.1 million deaths in 1990. About 68,000 of these occurred in children less than five years old. Almost all high-income countries have decreasing death rates, while the majority of low-income countries have increasing death rates due to traffic collisions. Middle-income countries have the highest rate with 20 deaths per 100,000 inhabitants, accounting for 80% of all road fatalities with 52% of all vehicles. While the death rate in Africa is the highest (24.1 per 100,000 inhabitants), the lowest rate is to be found in Europe (10.3 per 100,000 inhabitants).

Terminology

Traffic collisions can be classified by general types. Types of collision include head-on, road departure, rear-end, side collisions, and rollovers.

An Opel Vectra involved in a rollover crash.

Many different terms are commonly used to describe vehicle collisions. The World Health Organization uses the term road traffic injury, while the U.S. Census Bureau uses the term motor vehicle accidents (MVA), and Transport Canada uses the term "motor vehicle traffic collision" (MVTC). Other common terms include auto accident, car accident, car crash, car smash, car wreck, motor vehicle collision (MVC), personal injury collision (PIC), road accident, road traffic accident (RTA), road traffic collision (RTC), and road traffic incident (RTI) as well as more unofficial terms including smash-up, pile-up, and fender bender.

A rolled over box truck being handled by fire fighters in Jakarta, Indonesia.

Some organizations have begun to avoid the term "accident", instead preferring terms such as "collision", "crash" or "incident". This is because the term "accident" implies that there is no-one to blame, whereas most traffic collisions are the result of driving under the influence, excessive speed, distractions such as mobile phones or other risky behavior.

Historically, in the United States, the use of terms other than "accidents" had been criticized for holding back safety improvements, based on the idea that a culture of blame may discourage the involved parties from fully disclosing the facts, and thus frustrate attempts to address the real root causes.

Causes

A 1985 study by K. Rumar, using British and American crash reports as data, suggested 57% of crashes were due solely to driver factors, 27% to combined roadway and driver factors, 6% to combined vehicle and driver factors, 3% solely to roadway factors, 3% to combined roadway, driver, and vehicle factors, 2% solely to vehicle factors, and 1% to combined roadway and vehicle factors. Reducing the severity of injury in crashes is more important than reducing incidence and ranking incidence by broad categories of causes is misleading regarding severe injury reduction. Vehicle and road modifications are generally more effective than behavioral change efforts with the exception of certain laws such as required use of seat belts, motorcycle helmets and graduated licensing of teenagers.

Human Factors

Human factors in vehicle collisions include anything related to drivers and other road users that may contribute to a collision. Examples include driver behavior, visual and auditory acuity, decision-making ability, and reaction speed.

A 1985 report based on British and American crash data found driver error, intoxication and other human factors contribute wholly or partly to about 93% of crashes.

Drivers distracted by mobile devices had nearly four times greater risk of crashing their cars than those who were not. Dialing a phone is the most dangerous distraction, increasing a drivers' chance of crashing by 12 times, followed by reading or writing, which increased the risk by 10 times.

An RAC survey of British drivers found 78% of drivers thought they were highly skilled at driving, and most thought they were better than other drivers, a result suggesting overconfidence in their abilities. Nearly all drivers who had been in a crash did not believe themselves to be at fault. One survey of drivers reported that they thought the key elements of good driving were:

  * Controlling a car including a good awareness of the car's size and capabilities.

  * Reading and reacting to road conditions, weather, road signs and the environment.

  * Alertness, reading and anticipating the behavior of other drivers.

Although proficiency in these skills is taught and tested as part of the driving exam, a "good" driver can still be at a high risk of crashing because:

> "The feeling of being confident in more and more challenging situations is experienced as evidence of driving ability, and that 'proven' ability reinforces the feelings of confidence. Confidence feeds itself and grows unchecked until something happens – a near-miss or an accident".

An AXA survey concluded Irish drivers are very safety-conscious relative to other European drivers. However, this does not translate to significantly lower crash rates in Ireland.

Accompanying changes to road designs have been wide-scale adoptions of rules of the road alongside law enforcement policies that included drink-driving laws, setting of speed limits, and speed enforcement systems such as speed cameras. Some countries' driving tests have been expanded to test a new driver's behavior during emergencies, and their hazard perception.

There are demographic differences in crash rates. For example, although young people tend to have good reaction times, disproportionately more young male drivers feature in collisions, with researchers observing that many exhibit behaviors and attitudes to risk that can place them in more hazardous situations than other road users. This is reflected by actuaries when they set insurance rates for different age groups, partly based on their age, sex, and choice of vehicle. Older drivers with slower reactions might be expected to be involved in more collisions, but this has not been the case as they tend to drive less and, apparently, more cautiously. Attempts to impose traffic policies can be complicated by local circumstances and driver behavior. In 1969 Leeming warned that there is a balance to be struck when "improving" the safety of a road.

Conversely, a location that does not look dangerous may have a high crash frequency. This is, in part, because if drivers perceive a location as hazardous, they take more care. Collisions may be more likely to happen when hazardous road or traffic conditions are not obvious at a glance, or where the conditions are too complicated for the limited human machine to perceive and react in the time and distance available. High incidence of crashes is not indicative of high injury risk. Crashes are common in areas of high vehicle congestion, but fatal crashes occur disproportionately on rural roads at night when traffic is relatively light.

This phenomenon has been observed in risk compensation research, where the predicted reductions in collision rates have not occurred after legislative or technical changes. One study observed that the introduction of improved brakes resulted in more aggressive driving, and another argued that compulsory seat belt laws have not been accompanied by a clearly attributed fall in overall fatalities. Most claims of risk compensation offsetting the effects of vehicle regulation and belt use laws have been discredited by research using more refined data.

In the 1990s, Hans Monderman's studies of driver behavior led him to the realization that signs and regulations had an adverse effect on a driver's ability to interact safely with other road users. Monderman developed shared space principles, rooted in the principles of the woonerven of the 1970s. He concluded that the removal of highway clutter, while allowing drivers and other road users to mingle with equal priority, could help drivers recognize environmental clues. They relied on their cognitive skills alone, reducing traffic speeds radically and resulting in lower levels of road casualties and lower levels of congestion.

Some crashes are intended; staged crashes, for example, involve at least one party who hopes to crash a vehicle in order to submit lucrative claims to an insurance company. In the United States during the 1990s, criminals recruited Latin immigrants to deliberately crash cars, usually by cutting in front of another car and slamming on the brakes. It was an illegal and risky job, and they were typically paid only $100. Jose Luis Lopez Perez, a staged crash driver, died after one such maneuver, leading to an investigation that uncovered the increasing frequency of this type of crash.

Motor Vehicle Speed

The U.S. Department of Transportation's *Federal Highway Administration* review research on traffic speed in 1998. The summary says:

- The evidence shows the risk of having a crash is increased both for vehicles traveling slower than the average speed, and for those traveling above the average speed.

- The risk of being injured increases exponentially with speeds much faster than the median speed.

- The severity/lethality of a crash depends on the vehicle speed change at impact.

- There is limited evidence suggesting lower speed limits result in lower speeds on a system-wide basis.

- Most crashes related to speed involve speed too fast for the conditions.

- More research is needed to determine the effectiveness of traffic calming.

The Road and Traffic Authority (RTA) of the Australian state of New South Wales (NSW) asserts speeding (traveling too fast for the prevailing conditions or above the posted speed limit) is a factor in about 40 percent of road deaths. The RTA also say speeding increases the risk of a crash and its severity. On another web page, the RTA qualify their claims by referring to one specific piece of research from 1997, and writes "research has shown that the risk of a crash causing death or injury increases rapidly, even with small increases above an appropriately set speed limit."

The contributory factor report in the official British road casualty statistics show for 2006, that "exceeding speed limit" was a contributory factor in 5% of all casualty crashes (14% of all fatal crashes), and "traveling too fast for conditions" was a contributory factor in 11% of all casualty crashes (18% of all fatal crashes).

Assured Clear Distance Ahead

A common cause of collisions is driving faster than one can stop within their field of vision. Such practice is illegal and is particularly responsible for an increase of fatalities at night – when it occurs most.

Driver Impairment

Driver impairment describes factors that prevent the driver from driving at their normal level of skill. Common impairments include:

Alcohol

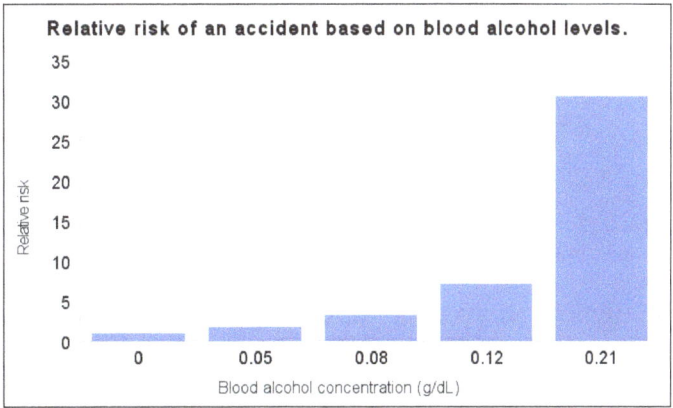

Relative risk of collisions based on blood alcohol levels.

According to the Government of Canada, coroner reports from 2008 suggested almost 40% of fatally injured drivers consumed some quantity of alcohol before the collision.

Physical Impairment

Poor eyesight and physical impairment, with many jurisdictions setting simple sight tests and requiring appropriate vehicle modifications before being allowed to drive.

Youth

Insurance statistics demonstrate a notably higher incidence of collisions and fatalities among drivers aged in their teens or early twenties, with insurance rates reflecting this data. These drivers have the highest incidence of both collisions and fatalities among all driver age groups, a fact that was observed well before the advent of mobile phones.

Females in this age group exhibit somewhat lower collision and fatality rates than males but still register well above the median for drivers of all ages. Also within this group, the highest collision incidence rate occurs within the first year of licensed driving. For this reason, many US states have enacted a zero-tolerance policy wherein receiving a moving violation within the first six months to one year of obtaining a license results in automatic license suspension. No US state allows fourteen year-olds to obtain drivers' licenses any longer.

Old Age

Old age, with some jurisdictions requiring driver retesting for reaction speed and eyesight after a certain age.

Sleep Deprivation

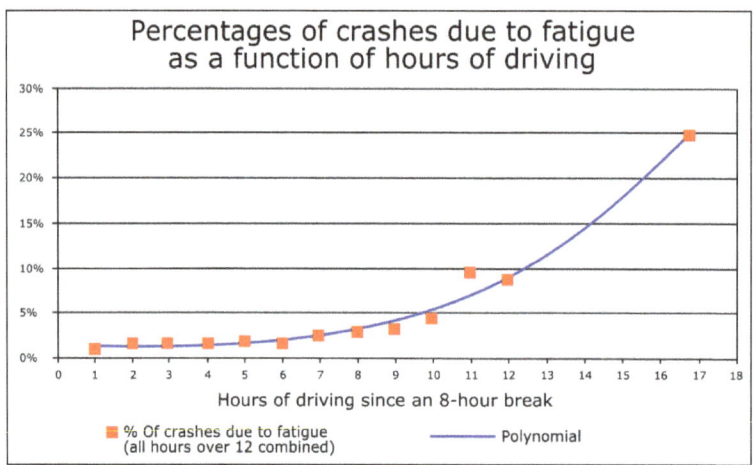

A graph outlining the relationship between number of hours driven and the percent of commercial truck crashes related to driver fatigue.

Various factors such as fatigue or sleep deprivation might increase the risk, or numbers of hours driving might increase the risk of an accident.

Drug Use

Including some prescription drugs, over the counter drugs (notably antihistamines, opioids and muscarinic antagonists), and illegal drugs.

Distraction

Research suggests that the driver's attention is affected by distracting sounds such as conversations and operating a mobile phone while driving. Many jurisdictions now restrict or outlaw the use of some types of phone within the car. Recent research conducted by British scientists suggests that music can also have an effect; classical music is considered to be calming, yet too much could relax the driver to a condition of distraction. On the other hand, hard rock may encourage the driver to step on the acceleration pedal, thus creating a potentially dangerous situation on the road.

Cell phone use is an increasingly significant problem on the roads and as the U.S. National Safety Council compiled more than 30 studies postulating that hands-free is not a safer option, because the brain remains distracted by the conversation and cannot focus solely on the task of driving.

Intent

Some traffic collisions are caused intentionally by a driver. For example, a collision may be caused by a driver who intends to commit suicide. Collisions may also be intentionally caused by people who hope to make an insurance claim against the other driver, or may be staged for such purposes as insurance fraud. Motor vehicles may also be involved in collisions as part of a deliberate effort to hurt other people, such as in a vehicle-ramming attack.

Combinations of Factors

Several conditions can combine to create a much worse situation, for example:

- Combining low doses of alcohol and cannabis has a more severe effect on driving performance than either cannabis or alcohol in isolation.

- Taking recommended doses of several drugs together, which individually do not cause impairment, may combine to bring on drowsiness or other impairment. This could be more pronounced in an elderly person whose renal function is less efficient than a younger person's.

Thus, there are situations when a person may be impaired, but still legally allowed to drive, and becomes a potential hazard to themselves and other road users. Pedestrians or cyclists are affected in the same way and can similarly jeopardize themselves or others when on the road.

Road Design

A 1985 US study showed that about 34% of serious crashes had contributing factors related to the roadway or its environment. Most of these crashes also involved a human factor. The road or environmental factor was either noted as making a significant contribution to the circumstances of the crash, or did not allow room to recover. In these circumstances, it is frequently the driver who is blamed rather than the road; those reporting the collisions have a tendency to overlook the human factors involved, such as the subtleties of design and maintenance that a driver could fail to observe or inadequately compensate for.

A potential long fall stopped by an early guardrail, ca. 1920. Guardrails, median barriers, or other physical objects can help reduce the consequences of a collision or minimize damage.

Research has shown that careful design and maintenance, with well-designed intersections, road surfaces, visibility and traffic control devices, can result in significant improvements in collision rates.

Individual roads also have widely differing performance in the event of an impact. In Europe, there are now EuroRAP tests that indicate how "self-explaining" and forgiving a particular road and its roadside would be in the event of a major incident.

In the UK, research has shown that investment in a safe road infrastructure program could yield a ⅓ reduction in road deaths, saving as much as £6 billion per year. A consortium of 13 major road safety stakeholders have formed the Campaign for Safe Road Design, which is calling on the UK Government to make safe road design a national transport priority.

Vehicle Design and Maintenance

A 2005 Chevrolet Malibu involved in a rollover crash.

Seat Belts

Research has shown that, across all collision types, it is less likely that seat belts were worn in collisions involving death or serious injury, rather than light injury; wearing a seat belt reduces the risk of death by about 45 percent. Seat belt use is controversial, with notable critics such as

Professor John Adams suggesting that their use may lead to a net increase in road casualties due to a phenomenon known as risk compensation. However, actual observation of driver behaviors before and after seat belt laws does not support the risk compensation hypothesis. Several important driving behaviors were observed on the road before and after the belt use law was enforced in Newfoundland, and in Nova Scotia during the same period without a law. Belt use increased from 16 percent to 77 percent in Newfoundland and remained virtually unchanged in Nova Scotia. Four driver behaviors (speed, stopping at intersections when the control light was amber, turning left in front of oncoming traffic, and gaps in following distance) were measured at various sites before and after the law. Changes in these behaviors in Newfoundland were similar to those in Nova Scotia, except that drivers in Newfoundland drove slower on expressways after the law, contrary to the risk compensation theory.

Maintenance

A well-designed and well-maintained vehicle, with good brakes, tires and well-adjusted suspension will be more controllable in an emergency and thus be better equipped to avoid collisions. Some mandatory vehicle inspection schemes include tests for some aspects of roadworthiness, such as the UK's MOT test or German TÜV conformance inspection.

The design of vehicles has also evolved to improve protection after collision, both for vehicle occupants and for those outside of the vehicle. Much of this work was led by automotive industry competition and technological innovation, leading to measures such as Saab's safety cage and reinforced roof pillars of 1946, Ford´s 1956 *Lifeguard* safety package, and Saab and Volvo's introduction of standard fit seatbelts in 1959. Other initiatives were accelerated as a reaction to consumer pressure, after publications such as Ralph Nader's 1965 book *Unsafe at Any Speed* accused motor manufacturers of indifference towards safety.

In the early 1970s, British Leyland started an intensive programme of vehicle safety research, producing a number of prototype experimental safety vehicles demonstrating various innovations for occupant and pedestrian protection such as air bags, anti-lock brakes, impact-absorbing side-panels, front and rear head restraints, run-flat tires, smooth and deformable front-ends, impact-absorbing bumpers, and retractable headlamps. Design has also been influenced by government legislation, such as the Euro NCAP impact test.

Common features designed to improve safety include thicker pillars, safety glass, interiors with no sharp edges, stronger bodies, other active or passive safety features, and smooth exteriors to reduce the consequences of an impact with pedestrians.

The UK Department for Transport publish road casualty statistics for each type of collision and vehicle through its Road Casualties Great Britain report. These statistics show a ten to one ratio of in-vehicle fatalities between types of car. In most cars, occupants have a 2–8% chance of death in a two-car collision.

Center of Gravity

Some crash types tend to have more serious consequences. Rollovers have become more common in recent years, perhaps due to increased popularity of taller SUVs, people carriers, and minivans, which have a higher center of gravity than standard passenger cars. Rollovers can be fatal, especially if the occupants are ejected because they were not wearing seat belts (83% of ejections during rollovers were fatal when the driver did not wear a seat belt, compared to 25% when they did). After a new design of Mercedes Benz notoriously failed a 'moose test' (sudden swerving to avoid an obstacle), some manufacturers enhanced suspension using stability control linked to an anti-lock braking system to reduce the likelihood of rollover. After retrofitting these systems to its models in 1999–2000, Mercedes saw its models involved in fewer crashes.

Now, about 40% of new US vehicles, mainly the SUVs, vans and pickup trucks that are more susceptible to rollover, are being produced with a lower center of gravity and enhanced suspension with stability control linked to its anti-lock braking system to reduce the risk of rollover and meet US federal requirements that mandate anti-rollover technology by September 2011.

Motorcycles

Motorcyclists have little protection other than their clothing and helmets. This difference is reflected in the casualty statistics, where they are more than twice as likely to suffer severely after a collision. In 2005, there were 198,735 road crashes with 271,017 reported casualties on roads in Great Britain. This included 3,201 deaths (1.1%) and 28,954 serious injuries (10.7%) overall. Of these casualties 178,302 (66%) were car users and 24,824 (9%) were motorcyclists, of whom 569 were killed (2.3%) and 5,939 seriously injured (24%).

Sociological Factors

Studies in United States have shown that poor people have a greater risk of dying in a car crash than people who are well-off. Car deaths are also higher in poorer states.

Similar studies in France have shown the same results. This may be due to working-class people having less access to secure equipment in cars, having older cars which are less protected against crash, and needing to cover more distance to go to work each day.

Other

Other possibly hazardous factors that may alter a driver's soundness on the road includes:

- Irritability.

- Following specifically distinct rules too bureaucratically, inflexibly or rigidly when unique circumstances might suggest otherwise.

- Sudden swerving into somebody's blind spot without first clearly making oneself visible through the wing mirror.

- Unfamiliarity with one's dashboard features, center console or other interior handling devices after a recent car purchase.

- Lack of visibility due to windshield design or sun glare.

- Distraction by scenery, a sexually attractive person or sexually suggestive advertising.

- Traffic safety culture, a variety of aspects of safety culture could impact on the number of crashes.

Prevention

A large body of knowledge has been amassed on how to prevent car crashes, and reduce the severity of those that do occur.

United Nations

Owing to the global and massive scale of the issue, with predictions that by 2020 road traffic deaths and injuries will exceed HIV/AIDS as a burden of death and disability, the United Nations and its subsidiary bodies have passed resolutions and held conferences on the issue. The first United Nations General Assembly resolution and debate was in 2003. The World Day of Remembrance for Road Traffic Victims was declared in 2005. In 2009 the first high level ministerial conference on road safety was held in Moscow.

The World Health Organization, a specialized agency of the United Nations Organization, in its Global Status Report on Road Safety 2009, estimates that over 90% of the world's fatalities on the roads occur in low-income and middle-income countries, which have only 48% of the world's registered vehicles, and predicts road traffic injuries will rise to become the fifth leading cause of death by 2030.

Collision Migration

Collisions migration refers to a situation where action to reduce road traffic collisions in one place may result in those collisions resurfacing elsewhere. For example, an accident blackspot may occur at a dangerous bend. The treatment for this may be to increase signage, post an advisory speed limit, apply a high-friction road surface, add crash barriers or any one of a number of other visible interventions. The immediate result may be to reduce collisions at the bend, but the subconscious relaxation on leaving the "dangerous" bend may cause drivers to act with fractionally less care on the rest of the road, resulting in an increase in collisions elsewhere on the road, and no overall improvement over the area. In the same way, increasing familiarity with the treated area will often result in a reduction over time to the previous level of care (regression to the mean) and may result in faster speeds around the bend due to perceived increased safety (risk compensation).

CRASHWORTHINESS

Crashworthiness is the ability of a structure to protect its occupants during an impact. This is commonly tested when investigating the safety of aircraft and vehicles. Depending on the nature of the impact and the vehicle involved, different criteria are used to determine the crashworthiness of the structure. Crashworthiness may be assessed either prospectively, using computer models (e.g., LS-DYNA, PAM-CRASH, MSC Dytran, MADYMO) or experiments, or retrospectively by analyzing crash outcomes. Several criteria are used to assess crashworthiness prospectively, including the deformation patterns of the vehicle structure, the acceleration experienced by the vehicle during an impact, and the probability of injury predicted by human body models. Injury probability is defined using criteria, which are mechanical parameters (e.g., force, acceleration, or deformation) that correlate with injury risk. A common injury criterion is the Head impact criterion (HIC). Crashworthiness is assessed retrospectively by analyzing injury risk in real-world crashes, often using regression or other statistical techniques to control for the myriad of confounders that are present in crashes.

CRASH TEST

A crash test is a form of destructive testing usually performed in order to ensure safe design standards in crashworthiness and crash compatibility for various modes of transportation or related systems and components.

Types

- Frontal-impact tests: Which is what most people initially think of when asked about a crash test. Vehicles usually impact a solid concrete wall at a specified speed, but these can also be vehicle impacting vehicle tests. SUVs have been singled out in these tests for a while, due to the high ride-height that they often have.

- Moderate Overlap tests: In which only part of the front of the car impacts with a barrier (vehicle). These are important, as impact forces (approximately) remain the same as with a frontal impact test, but a smaller fraction of the car is required to absorb all of the force. These tests are often realized by cars turning into oncoming traffic. This type of testing is done by the U.S.A. Insurance Institute for Highway Safety (IIHS), EuroNCAP, Australasian New Car Assessment Program (ANCAP) and ASEAN NCAP.

- Small Overlap tests: This is where only a small portion of the car's structure strikes an object such as a pole or a tree, or if a car were to clip another car. This is the most demanding test because it loads the most force onto the structure of the car at any given speed. These are usually conducted at 15-20% of the front vehicle structure.

- Side-impact tests: These forms of accidents have a very significant likelihood of fatality, as cars do not have a significant crumple zone to absorb the impact forces before an occupant is injured.

- Poll-impact tests: A difficult test which places a large amount of force on a small proportion on the side of the vehicle.

- Roll-over tests: Which tests a car's ability (specifically the pillars holding the roof) to support itself in a dynamic impact. More recently, dynamic rollover tests have been proposed in lieu of static crush testing (video).

- Roadside hardware crash tests: Are used to ensure crash barriers and crash cushions will protect vehicle occupants from roadside hazards, and also to ensure that guard rails, sign posts, light poles and similar appurtenances do not pose an undue hazard to vehicle occupants.

- Old versus new: Often an old and big car against a small and new car, or two different generations of the same car model. These tests are performed to show the advancements in crash-worthiness.

- Computer model: Because of the cost of full-scale crash tests, engineers often run many simulated crash tests using computer models to refine their vehicle or barrier designs before conducting live tests.

2016 Honda Fit striking a wall head-on at 56 km/h.

- Sled testing: A cost-effective way of testing components such as airbags and seat belts is conducting sled crash testing. The two most common types of sled systems are reverse-firing sleds which are fired from a standstill, and decelerating sleds which are accelerated from a starting point and stopped in the crash area with a hydraulic ram. It can also be used to evaluate the whiplash protection of a vehicle's seat.

Frontal small-overlap crash test of a 2012 Honda Odyssey.

GLOBAL NCAP CRASH TEST

NCAP stands for New Car Assessment Program. To give a brief background, in 1978 USA became the first country to come up with a programme to provide car crashworthiness information to consumers, which eventually expanded to crash testing and reporting the results. The US-NCAP model formed the basis for similar programmes in other regions, and today there is the Australasian NCAP, Euro NCAP, Japan NCAP, ASEAN NCAP, China NCAP, Korean NCAP and Latin NCAP. Global NCAP, an independent charity registered in the UK, was formed in 2011 to enhance cooperation between the various NCAPs and primarily promote vehicle crash-testing and reporting in emerging markets. 'Safer Cars For India' and 'Safer Cars For Africa' are its key initiatives at the moment.

Crash Test Procedure through Global NCAP

Every NCAP has its own protocol to crash-test and score cars, and so the results are not interchangeable. Euro NCAP, for instance, conducts full frontal, front offset, side impact and side pole tests. Global NCAP ratings, on the other hand, are based on front offset crash tests alone. A front offset crash test is designed to simulate a head-on collision between two cars. In the Global NCAP test, the car is driven at 64kph and with 40 percent overlap into a deformable barrier which is the equivalent of a crash between two cars of the same weight, both moving at 50kph.

Procedure of Giving Crash Score to Cars through Global NCAP

Technically, a car cannot 'fail' an NCAP test, as it can a government regulation test. Each car under the NCAP, as you may be familiar, is given a rating on a 5-star scale – the higher the star rating, the safer the car. The rating itself is based on the Adult Occupant Protection and Child Occupant Protection scores resulting from the crash test. These scores are primarily derived from readings of the crash-test dummies but additional points may be awarded for the presence of certain safety features. Additionally, Global NCAP mandates a driver's side airbag as the minimum requirement to qualify for a one star rating. This should explain why non-airbag versions of the Tata Zest and Volkswagen Polo received zero stars, while airbag-equipped versions tested later were rated 4-star cars. In time, more requirements will be introduced too. For instance, ESC could become a minimum star requirement in the tests on Indian cars in the years to come.

The 17-point Adult Occupant Protection score takes into account driver injury readings from four body regions – head and neck; chest; knee, femur and pelvis, and leg and foot. An additional point is given to cars with a seatbelt reminder, four-channel ABS and some form of side-impact protection, tested by a relevant authority.

The primary basis for the 49-point Child Occupant Protection score is readings from the 18-month-old and 3-year-old-sized dummies placed in manufacturer recommended child seats. Additional points are given for child restraint system markings, provision of three point seat belts, Isofix, etc.

Procedure of Selecting Cars through Global NCAP

The cars to be tested are bought by the agency from a showroom. Global NCAP uses cars in base trim for the test and the idea is to establish a baseline level of safety a buyer gets even on the most affordable version of a car. Carmakers, however, are allowed and encouraged to send an improved or higher-spec car with more safety features for an additional crash test and rating too. The Tata Zest, Volkswagen Polo and Honda Mobilio have been tested twice, while the Renault Kwid has been tested four times since 2016. In the event a manufacturer is providing a test car, the model is selected straight off the assembly line by Global NCAP as per a strict internal protocol.

Car Crash Test rating

As things stand, not any time soon. Global NCAP, as an independent body, has limited resources to conduct crash tests on each and every car. However, the broad goal is to create awareness among buyers on safe cars that would translate into a demand for a safety rating. Just for reference, USA mandates a crash worthiness rating label on cars for sale. On the other hand, an NCAP rating is not mandatory in Europe but its absence is viewed as suspect by buyers, explaining why 80 percent of new models are sent for a Euro NCAP rating by the manufacturers themselves. A good safety rating is good for business.

References

- One in a Million: Toledo Assembly Complex Marks Production of One-Millionth Jeep Wrangler JK" (Press release). Chrysler Group LLC. Retrieved 16 February 2014

- Automobile-safety-system-seminar-report, seminar: seminarstopics.com, Retrieved 28 July, 2019

- Rudolphi, Josie M.; Campo, Shelly; Gerr, Fred; Rohlman, Diane S. (May 2018). "Social and Individual Influences on Tractor Operating Practices of Young Adult Agricultural Workers". Journal of Adolescent Health. 62 (5): 605–611. Doi:10.1016/j.jadohealth.2017.11.300. ISSN 1054-139X

- Breaking-down-the-global-ncap-crash-test, car-news: autocarindia.com, Retrieved 29 August, 2019

- Crosby, Rachel (18 September 2017). "24 face fraud charges in Las Vegas car crash scheme". Las Vegas Review-Journal. Retrieved 3 December 2017

- "Consumers Claim Tires Are Defective, File Class-Action Suit Against Goodyear". Wall Street Journal. 24 November 2000. Retrieved 31 October 2017

INDEX